Compendium of Plant Genomes

Series Editor

Chittaranjan Kole, Raja Ramanna Fellow, Government of India, ICAR-National Research Center on Plant Biotechnology, Pusa, New Delhi, India

Whole-genome sequencing is at the cutting edge of life sciences in the new millennium. Since the first genome sequencing of the model plant Arabidopsis thaliana in 2000, whole genomes of about 100 plant species have been sequenced and genome sequences of several other plants are in the pipeline. Research publications on these genome initiatives are scattered on dedicated web sites and in journals with all too brief descriptions. The individual volumes elucidate the background history of the national and international genome initiatives; public and private partners involved; strategies and genomic resources and tools utilized; enumeration on the sequences and their assembly; repetitive sequences; gene annotation and genome duplication. In addition, synteny with other sequences, comparison of gene families and most importantly potential of the genome sequence information for gene pool characterization and genetic improvement of crop plants are described.

Interested in editing a volume on a crop or model plant?
Please contact Prof. C. Kole, Series Editor, at ckoleorg@gmail.com

More information about this series at http://www.springer.com/series/11805

N. Manikanda Boopathi •
M. Raveendran • Chittaranjan Kole
Editors

The Moringa Genome

Springer

Editors
N. Manikanda Boopathi
Department of Plant Biotechnology
Tamil Nadu Agricultural University
Coimbatore, Tamil Nadu, India

M. Raveendran
Department of Plant Biotechnology
Tamil Nadu Agricultural University
Coimbatore, Tamil Nadu, India

Chittaranjan Kole
ICAR-National Institute for Plant
Biotechnology
New Delhi, India

ISSN 2199-4781 ISSN 2199-479X (electronic)
Compendium of Plant Genomes
ISBN 978-3-030-80955-3 ISBN 978-3-030-80956-0 (eBook)
https://doi.org/10.1007/978-3-030-80956-0

This Springer imprint is published by the registered company Springer Nature Switzerland AG
The registered company address is: Gewerbestrasse 11, 6330 Cham, Switzerland

This book series is dedicated to my wife Phullara and our children Sourav and Devleena

Chittaranjan Kole

This book is dedicated to all the forefathers who have preserved the Moringa spp., for the future.

Preface to the Series

Genome sequencing has emerged as the leading discipline in the plant sciences coinciding with the start of the new century. For much of the twentieth century, plant geneticists were only successful in delineating putative chromosomal location, function, and changes in genes indirectly through the use of a number of "markers" physically linked to them. These included visible or morphological, cytological, protein, and molecular or DNA markers. Among them, the first DNA marker, the RFLPs, introduced a revolutionary change in plant genetics and breeding in the mid-1980s, mainly because of their infinite number and thus potential to cover maximum chromosomal regions, phenotypic neutrality, absence of epistasis, and codominant nature. An array of other hybridization-based markers, PCR-based markers, and markers based on both facilitated construction of genetic linkage maps, mapping of genes controlling simply inherited traits, and even gene clusters (QTLs) controlling polygenic traits in a large number of model and crop plants. During this period, a number of new mapping populations beyond F2 were utilized and a number of computer programs were developed for map construction, mapping of genes, and for mapping of polygenic clusters or QTLs. Molecular markers were also used in the studies of evolution and phylogenetic relationship, genetic diversity, DNA fingerprinting, and map-based cloning. Markers tightly linked to the genes were used in crop improvement employing the so-called marker-assisted selection. These strategies of molecular genetic mapping and molecular breeding made a spectacular impact during the last one and a half decades of the twentieth century. But still, they remained "indirect" approaches for elucidation and utilization of plant genomes since much of the chromosomes remained unknown and the complete chemical depiction of them was yet to be unraveled.

Physical mapping of genomes was the obvious consequence that facilitated the development of the "genomic resources" including BAC and YAC libraries to develop physical maps in some plant genomes. Subsequently, integrated genetic–physical maps were also developed in many plants. This led to the concept of structural genomics. Later on, emphasis was laid on EST and transcriptome analysis to decipher the function of the active gene sequences leading to another concept defined as functional genomics. The advent of techniques of bacteriophage gene and DNA sequencing in the 1970s was extended to facilitate sequencing of these genomic resources in the past decade of the twentieth century.

As expected, sequencing of chromosomal regions would have led to too much data to store, characterize, and utilize with the then available computer software could handle. But the development of information technology made the life of biologists easier by leading to a swift and sweet marriage of biology and informatics, and a new subject was born—bioinformatics.

Thus, the evolution of the concepts, strategies, and tools of sequencing and bioinformatics reinforced the subject of genomics—structural and functional. Today, genome sequencing has traveled much beyond biology and involves biophysics, biochemistry, and bioinformatics!

Thanks to the efforts of both public and private agencies, genome sequencing strategies are evolving very fast, leading to cheaper, quicker, and automated techniques right from clone-by-clone and whole-genome shotgun approaches to a succession of second-generation sequencing methods. The development of software of different generations facilitated this genome sequencing. At the same time, newer concepts and strategies were emerging to handle sequencing of the complex genomes, particularly the polyploids.

It became a reality to chemically—and so directly—define plant genomes, popularly called whole-genome sequencing or simply genome sequencing.

The history of plant genome sequencing will always cite the sequencing of the genome of the model plant Arabidopsis thaliana in 2000 that was followed by sequencing the genome of the crop and model plant rice in 2002. Since then, the number of sequenced genomes of higher plants has been increasing exponentially, mainly due to the development of cheaper and quicker genomic techniques and, most importantly, the development of collaborative platforms such as national and international consortia involving partners from public and/or private agencies.

As I write this preface for the first volume of the new series "Compendium of Plant Genomes," a net search tells me that complete or nearly complete whole-genome sequencing of 45 crop plants, 8 crop and model plants, 8 model plants, 15 crop progenitors and relatives, and 3 basal plants is accomplished, the majority of which are in the public domain. This means that we nowadays know many of our model and crop plants chemically, i.e., directly, and we may depict them and utilize them precisely better than ever. Genome sequencing has covered all groups of crop plants. Hence, information on the precise depiction of plant genomes and the scope of their utilization are growing rapidly every day. However, the information is scattered in research articles and review papers in journals and dedicated Web pages of the consortia and databases. There is no compilation of plant genomes and the opportunity of using the information in sequence-assisted breeding or further genomic studies. This is the underlying rationale for starting this book series, with each volume dedicated to a particular plant.

Plant genome science has emerged as an important subject in academia, and the present compendium of plant genomes will be highly useful to both students and teaching faculties. Most importantly, research scientists involved in genomics research will have access to systematic deliberations on the plant genomes of their interest. Elucidation of plant genomes is of interest not only for the geneticists and breeders, but also for practitioners of an array of plant science disciplines, such as taxonomy, evolution, cytology, physiology, pathology, entomology, nematology, crop production, biochemistry, and obviously bioinformatics. It must be mentioned that information regarding each plant genome is ever-growing. The contents of the volumes of this compendium are, therefore, focusing on the basic aspects of the genomes and their utility. They include information on the academic and/or economic importance of the plants, a description of their genomes from a molecular genetic and cytogenetic point of view, and the genomic resources developed. Detailed deliberations focus on the background history of the national and international genome initiatives, public and private partners involved, strategies and genomic resources and tools utilized, enumeration on the sequences and their assembly, repetitive sequences, gene annotation, and genome duplication. In addition, synteny with other sequences, comparison of gene families, and, most importantly, the potential of the genome sequence information for gene pool characterization through genotyping by sequencing (GBS) and genetic improvement of crop plants have been described. As expected, there is a lot of variation of these topics in the volumes based on the information available on the crop, model, or reference plants.

I must confess that as the series editor, it has been a daunting task for me to work on such a huge and broad knowledge base that spans so many diverse plant species. However, pioneering scientists with lifetime experience and expertise on the particular crops did excellent jobs editing the respective volumes. I myself have been a small science worker on plant genomes since the mid-1980s and that provided me the opportunity to personally know several stalwarts of plant genomics from all over the globe. Most, if not all, of the volume editors are my longtime friends and colleagues. It has been highly comfortable and enriching for me to work with them on this book series. To be honest, while working on this series I have been and will remain a student first, a science worker second, and a series editor last. And I must express my gratitude to the volume editors and the chapter authors for providing me the opportunity to work with them on this compendium.

I also wish to mention here my thanks and gratitude to the Springer staff, particularly Dr. Christina Eckey and Dr. Jutta Lindenborn for the earlier set of volumes and presently Ing. Zuzana Bernhart for all their timely help and support.

I always had to set aside additional hours to edit books besides my professional and personal commitments—hours I could and should have given to my wife, Phullara, and our kids, Sourav and Devleena. I must mention that they not only allowed me the freedom to take away those hours from them, but also offered their support in the editing job itself. I am really not sure whether my dedication of this compendium to them will suffice to do justice to their sacrifices for the interest of science and the science community.

New Delhi, India Chittaranjan Kole

Preface

Moringa has long been used as a vegetable and medicine, and traditionally, people have expressed the benefits of Moringa based on their real-life experiences.

Moringa leaves are certainly worth paying special attention. Conventional medications have used these leaves over centuries in several countries to cure a range of diseases/disorders. Recent clinical experiments have also endorsed these treatments.

Further, nutritional analysis has shown that the leaves are remarkably high in protein, more importantly they contain all of the essential amino acids, including the ones that are required for lactating mothers and infants. Therefore, it is considered as rarest of rare plant food. Moringa also has a package of all essential vitamins and minerals and it is widely employed as an affordable food supplement to combat malnutrition and related diseases.

Above all, an agricultural production point of view, Moringa is a fast-growing, drought-resistant tree that can be grown in a resource-limited environment and even in barren lands.

Hence, it is no wonder to name it a miracle tree.

This book compiles all the required information on Moringa in a comprehensive style, in such a way that it will serve as a resource guide for those who are interested to evolve improved cultivars for the given environment.

Coimbatore, India — N. Manikanda Boopathi
Coimbatore, India — M. Raveendran
New Delhi, India — Chittaranjan Kole
January 2021

Contents

Contributors

Bashir Yusuf Abubakar Department of Botany, Ahmadu Bello University, Zaria, Nigeria

S. Amutha Community Science College and Research Institute, Tamil Nadu Agricultural University, Madurai, India

K. A. Apoorva Department of Biotechnology and Crop Improvement, College of Horticulture, University of Horticultural Sciences, Bengaluru, India

R. Balakumbahan Department of Vegetable Crops, Horticultural College and Research Institute, Tamil Nadu Agricultural University (TNAU), Periyakulam, India

A. Bharathi Department of Plant Biotechnology, Centre for Plant Molecular Biology & Biotechnology, Tamil Nadu Agricultural University, Coimbatore, India

N. Manikanda Boopathi Department of Plant Biotechnology, Center for Plant Molecular Biology & Biotechnology, Tamil Nadu Agricultural University, Coimbatore, India

J. Deepa Department of Plant Biotechnology, Centre for Plant Molecular Biology & Biotechnology, Tamil Nadu Agricultural University, Coimbatore, India

J. Devi Priya Community Science College and Research Institute, Tamil Nadu Agricultural University, Madurai, India

M. Doddamani Department of Biotechnology and Crop Improvement, College of Horticulture, University of Horticultural Sciences, Bengaluru, India

K. Elango Horticultural Research Station, Tamil Nadu Agricultural University, Kodaikanal, India;
Department of Agricultural Entomology, Tamil Nadu Agricultural University, Coimbatore, India

B. Fakrudin Department of Biotechnology and Crop Improvement, College of Horticulture, University of Horticultural Sciences, Bengaluru, India

K. Govindaraju Centre for Ocean Research (DST-FIST Sponsored Centre), Sathyabama Institute of Science and Technology, Chennai, India

J. D. Harshith Department of Plant Biotechnology, Center for Plant Molecular Biology & Biotechnology, Tamil Nadu Agricultural University, Coimbatore, India

G. Hemalatha Community Science College and Research Institute, Tamil Nadu Agricultural University, Madurai, India

R. C. Jagadeesha Department of Biotechnology and Crop Improvement, College of Horticulture, University of Horticultural Sciences, Bengaluru, India

K. Jothilakshmi Community Science College and Research Institute, Tamil Nadu Agricultural University, Madurai, India

S. Kadam Department of Biotechnology and Crop Improvement, College of Horticulture, University of Horticultural Sciences, Bengaluru, India

M. Kalyanasundaram Department of Nano Science and Technology, Tamil Nadu Agricultural University, Coimbatore, India

Rohit Kambale Department of Plant Biotechnology, Centre for Plant Molecular Biology & Biotechnology, Tamil Nadu Agricultural University, Coimbatore, India

M. Kannan Department of Nano Science and Technology, Tamil Nadu Agricultural University, Coimbatore, India

Chittaranjan Kole Raja Ramanna Fellow, ICAR-National Institute for Plant Biotechnology, New Delhi, India

A. Lakshmanan Department of Nano Science and Technology, Tamil Nadu Agricultural University, Coimbatore, India

T. N. Lakshmidevamma Department of Biotechnology and Crop Improvement, College of Horticulture, University of Horticultural Sciences, Bengaluru, India

N. Manavalan Government Yoga and Naturopathy Medical College, Chennai, India

Jayakanthan Mannu Department of Plant Molecular Biology and Bioinformatics, Centre for Plant Molecular Biology & Biotechnology, Tamil Nadu Agricultural University, Coimbatore, India

Saranya Nallusamy Department of Plant Molecular Biology and Bioinformatics, Centre for Plant Molecular Biology & Biotechnology, Tamil Nadu Agricultural University, Coimbatore, India

R. S. Nayana Department of Biotechnology and Crop Improvement, College of Horticulture, University of Horticultural Sciences, Bengaluru, India

M. Raveendran Department of Plant Biotechnology, Center for Plant Molecular Biology & Biotechnology, Tamil Nadu Agricultural University, Coimbatore, India

V. P. Santhanakrishnan Department of Plant Biotechnology, Center for Plant Molecular Biology & Biotechnology, Tamil Nadu Agricultural University, Coimbatore, India

S. P. Gautham Suresh Department of Biotechnology and Crop Improvement, College of Horticulture, University of Horticultural Sciences, Bengaluru, India

J. Ugalat Department of Biotechnology and Crop Improvement, College of Horticulture, University of Horticultural Sciences, Bengaluru, India

R. Veera Ranjani Department of Plant Biotechnology, Center for Plant Molecular Biology & Biotechnlogy, Tamil Nadu Agricultural University, Coimbatore, India

M. Williams Department of Plant Biotechnology, Centre for Plant Molecular Biology & Biotechnology, Tamil Nadu Agricultural University, Coimbatore, India

R. Yogeshwari Community Science College and Research Institute, Tamil Nadu Agricultural University, Madurai, India

Abbreviations

18S rDNA	18 S Ribosomal DNA
AFLP	Amplified Fragment Length Polymorphism
AMOVA	Analysis Of Molecular Variance
atpB	ATP Synthase Beta Subunit
AVRDC	Asian Vegetable Research And Development Center
BA	Benzyl Adenine
BAC	Bacterial Artificial Chromosomes
BAP	6-Benzylaminopurine
BARC	Baba Atomic Research Center
BLAST	Basic Local Alignment Search Tool
BOLD	Barcode Of Life Data Systems
BUSCO	Benchmarking Universal Single-Copy Orthologs
CAPS	Cleaved Amplified Polymorphic Sequences
CBoL	Consortium For The Barcode Of Life
cDNA	Complementary DNA
CDS	Coding Sequences
CHIAS	Chromosome Image Analyzing System
CITES	Convention On International Trade In Endangered Species Of Wild Fauna And Flora
cMC	Centimcclintock
CMS	Cytoplasmic Male Sterility
CNVs	Copy Number Variations
COI	Cytochrome C Oxidase Subunit 1
COVID-19	Corona Virus Disease 2019
CRISPR/Cas	Clustered Regularly Interspaced Short Palindromic Repeats/CRISPR-Associated Protein
CTAB	Cetyl Trimethylammonium Bromide
DBG	De Bruijn Graph
DEGs	Differentially Expressed Genes
DH	Doubled Haploids
DNA	Deoxyribonucleic Acid
EDII	Entrepreneurship Development And Innovation Institute
EST	Expressed Sequence Tags
Fe	Iron
FISH	Fluorescence In Situ Hybridization

GISH	Genomic In Situ Hybridization
GO	Gene Ontology
GWAS	Genome-Wide Association Studies
HDL	High-Density Lipoprotein
Hi-C	High Throughput Chromatin Conformation Capture
HPLC	High-Performance Liquid Chromatography
IAA	Indole-3-Acetic Acid
IBA	Indole-3-Butyric Acid
IDA	Iron Deficiency Anaemia
IDF	International Diabetes Federation
IMGC	International Moringa Germplasm Collection
INDELS	Insertion-Deletions
IR	Inverted Repeat
IRAP	Inter-Retrotransposon Amplified Polymorphism
ISSRs	Inter Simple Sequence Repeats
ITS	Internal Transcribed Spacer (Consisting Of ITS1 And ITS2)
IYFV	International Year Of Fruits And Vegetables
KEGG	Kyoto Encyclopedia Of Genes And Genomes
LC/MS	Liquid Chromatography/Mass Spectrophotometry
LDL	Low-Density Lipoprotein
LSC	Large Single Copy
MAS	Marker Assisted Selection
matK	Maturase K
Mb	Mega Bases
McFISH	Multi-Colour FISH
MEMO	Methanolic Extract of Moringa Oleifera
M-FISH	Metaphase FISH
MFPP	Micronized Ferric Pyrophosphate
miRNA	MicroRNA
MISA	Microsatellite Identification Tool
ML	Maximum Likelihood
MS Medium	Murashige and Skoog Basal Medium
MVs	Microvesicles
NAA	1-Naphthaleneacetic Acid
NCBI	National Center For Biotechnology Information
NGO	Non-Government Organization
NGS	Next-Generation Sequencing
OLC	Overlap Layout Consensus
ONT	Oxford Nanopore Technologies
PAVs	Presence/Absence Variations
PCA	Principal Coordinate Analysis
PCR	Polymerase Chain Reaction
PEG	Polyethyleneglycol
PFGE	Pulse Field Gel Electrophoresis
pg	Pico Gram
PGRs	Plant Growth Regulators

PIC	Polymorphic Information Content
PKM	Periyakulam
ppm	Parts Per Million
PTC	Plant Tissue Culture
qRT-PCR	Quantitative Real Time PCR
QTL	Quantitative Trait Loci
RAMP	Randomly Amplified Microsatellite Polymorphism
RAPD	Random Amplified Polymorphic DNA
rbcL	Ribulose-1,5-Bisphosphate Carboxylase Large Subunit
RBIP	Retrotransposon-Based Insertion Polymorphism
RDA	Recommended Dietary Allowances
rDNA	Recombinant DNA
REMAP	Retrotransposon-Microsatellite Amplified Polymorphism
RFLP	Restriction Fragment Length Polymorphism
RNA	Ribonucleic Acid
ROS	Reactive Oxygen Species
RSM	Response Surface Methodology
RT-PCR	Reverse Transcription Polymerase Chain Reaction
SCAR	Sequence Characterized Amplified Regions
SCOT	Start Codon Targeted Marker
smRNA-FISH	Single-Molecule RNA-FISH
SMRT	Single Molecule Real Time
SNP	Single Nucleotide Polymorphism
SRAP	Sequence-Related Amplified Polymorphism
SSC	Small Single Copy
SSCP	Single-Strand Conformation Polymorphism
SSDs	Small-Scale Duplications
SSR	Simple Sequence Repeat
TE	Transposable Elements
TGS	Third-Generation Sequencing
TRAP	Target Region Amplification Polymorphism
UN-FAO	United Nation's-Food And Agriculture Organization
UNAM	Universidad Nacional Autónoma De México
UPLC-ESI-MS/MS	Ultra-Performance Liquid Chromatography-Electrospray Tandem Mass Spectrometry/Mass Spectrometry
v/v	Volume By Volume
w/v	Weight By Volume
WGD	Whole Genome Duplication
WGS	Whole-Genome Shotgun Sequencing
YAC	Yeast Artificial Chromosomes
ZMW	Zero-Mode Waveguide

1 Moringa and Its Importance

N. Manikanda Boopathi and M. Raveendran

Abstract

In the recent past, *Moringa oleifera* Lam (Moringaceae) has been recognized as an economically and nutritionally important crop owing to its health benefits. Moringa is widespread and found in all tropical and subtropical climates. As this crop is grown by different groups of people, it is known by several regional names such as drumstick tree, sajiwan, kelor, murungai, marango, mlonge, mulangay, saijhan, ben oil tree and sajna. Moringa is valued for its high nutritional value and medicinally important phytochemicals. Edible parts of this plant are found to contain necessary nutritional compounds such as proteins, essential and non-essential amino acids, vitamins, minerals, antioxidants and other phenolic compounds. They have been historically consumed over centuries and have been employed in the indigenous system of medicine for the treatment of different maladies or disorders. Leaves, roots, seed, bark, fruit, flowers and immature pods of Moringa were found to have antioxidant, antidiabetic, antibacterial, antifungal, antitumour, anti-inflammatory, antiulcer, antispasmodic, diuretic antihypertensive, hepatoprotective, antipyretic, antiepileptic, cardiac and cholesterol-lowering activities. This chapter focuses on the nutritional and medicinal values of Moringa, their application in medicine along with their pharmacological properties besides providing an overview of promising cultivars and their cultivation.

1.1 Introduction

Universally, policy makers and scientists are always looking for alternative crops and strategies to increase food productions as the productivity of agricultural and horticultural crops is threatened by continuous climate change. It is imperative to double the production by 2050 to ensure the food and energy supply to the growing world population (Ray et al. 2013).

The FAO has projected that between 2005 and 2050, the area under cropland must be increased to a minimum of 69 million ha (Alexandratos and Bruinsma 2012) to meet the food demand in 2050. Such effort is considered as a gigantic challenge, owing to unpredictable weather conditions (including erratic precipitations and temperatures besides alterations in CO_2 levels; Gruda et al. 2019). This implies that we need to look for alternative crops, which can thrive well in these unpredictable harsh environments.

Moreover, ~95% of the human calories are supplied by only 30 crop species, of which wheat, maize, and rice are considered as staple foods.

N. M. Boopathi (✉) · M. Raveendran
Department of Plant Biotechnology, Center for Plant Molecular Biology and Biotechnology, Tamil Nadu Agricultural University, Coimbatore, India
e-mail: nmboopathi@tnau.ac.in

N. M. Boopathi et al. (eds.), *The Moringa Genome*, Compendium of Plant Genomes,
https://doi.org/10.1007/978-3-030-80956-0_1

History (such as the Bengal famine due to the failure of rice crops and potato famine caused by late blight) has indicated that depending on few crops may have a serious threat to food security. Therefore, to provide decent lives for all on the healthy planet by 2030 (as outlined in the United Nations' 17 Sustainable Development Goals; https://www.un.org/sustainabledevelopment/), it is essential to diversify and stabilize the global food supply, increase productivity and handle the malnutrition issues with affordable strategies.

Undernutrition, the long-pending health issue, affects almost 25% of children living on the earth under the age of five. Juvenile undernutrition also results in long-term health shortfalls; besides, it causes lessened cerebral potential and elevated chronic disease risks and is the reason for nearly half of children's fatalities. Among the several factors that cause undernutrition, food insecurity and lack of access to health care are the key players. Consequently, it is widely accepted that dietary intervention is an affordable strategy to systematically diminish the ill effects of undernutrition (Rouhani et al. 2020).

Consequently, to tackle the undernutrition, food aid programs with manufactured nutrient supplementation (such as nutrient-rich powders, high-energy food sources, and vitamin A drops) have been designed in developing and underdeveloped nations. On the other hand, such programs are found to be costlier and involve huge manpower to handle these programs.

The UN General Assembly designated the year 2021, as the International Year of Fruits and Vegetables (IYFV) by highlighting health and nutritional benefits through sustainable production and provide a chance to promote the role of fruits and vegetables in human nutrition, food security, and health besides reaching UN Sustainable Development Goals.

To this end, evaluation of local and affordable resources should be focused and explored for expanding nutritional security as more focus is put on regional food-based methodologies. At this juncture, it is worth mentioning that Moringa (*Moringa oleifera* Lam., Family *Moringaceae*) can be used as an affordable plant nutrient source that can be grown in resource-limited environments of tropical and subtropical countries.

Moringa is referred to in different names as listed in Table 1.1 and widely used at the personal and communal level as it provides affordable nutritionally rich food supplements.

It is believed that by promoting appropriate educational strategies which explain uses and nutritional gains that can be obtained from Moringa, the global nutritional hunger can be reduced to half and the malnutrition-related maladies found in global communities could be avoided. This has already been established in the regions that are affected by insufficient nutrition (such as Africa, Asia, Latin America and the Caribbean). People living in these resource-limited environments have affordably grown and utilized *M. oleifera* and relieved from undernourishment (Palada 1996).

Moringa spp., is a tropical perennial vegetable crop of Indian and African origin. In addition to *Moringa oleifera*, 12 species of *Moringa* are available in the world (Table 1.2). Among the thirteen species of *Moringa,* two are available in the Indian sub-continent and the remaining eleven species are found in the African continent.

Among the 13 *Moringa* species, *M. oleifera* is the only species that is commercially utilized for cultivation and human consumption. Though Moringa is predominantly cultivated in India and Africa, now it is being grown as a noteworthy economic and nutritional crop in Egypt, the Philippines, Malaysia, Singapore, Burma, Sri Lanka, Thailand, Myanmar, Pakistan, Cuba, Jamaica, Nigeria and West Indian Islands. Moringa cultivation in India, especially in the Southern part of India, is having long folklore practice for its pod and is used as a routine vegetable in south Indian culinary preparations. Owing to the nutrient richness and been called as a super food, moringa cultivation for leaves have now gained popularity among Indian farmers and other aforesaid countries.

The United Nations (UN)—Food and Agriculture Organization (FAO) has promoted Moringa as a water stress tolerant but quickly growing plant that has a significant amount of

Table 1.1 Different names that are used to denote *Moringa oleifera* in different languages

Language	Name used for *Moringa oleifera*
Latin	*Moringa oleifera*
Tamil	Murungai
Sanskrit	Subhanjana
Hindi	Sainjana, Saguna
Punjabi	Soanjana, Sainjana
Bengali	Sojne danta
Gujarati	Suragavo
Marathi	Shevga
Telugu	Munaga, Mulaga,
Malayalam	Sigru, Muringa
Ayurveda	Raktaka, Haritashaaka Tikshnagandhaa and Akshiva
Unani	Sahajan
Arabian	Rawag
French	Morungue, Moringe à graine ailée
Spanish	Ben oil tree, Moringa, Àngela,
Portuguese	Moringueiro, Moringa,
Chinese	La Ken
English	Horseradish/Drumstick/Ben tree

Table 1.2 Geographical location of different *Moringa* species

Species	Geographical distribution of *Moringa* spp.,
Trees with Slender Stem	
M. concanensis Nimmo and *M. oleifera* Lam.	India
M. peregrina (Forssk) Fiori	Fiori Red Sea, Arabia, Horn of Africa
Trees with Bottle like Stem	
M. drouhardii Jum. and *M. hildebrandtii* Engl.	Madagascar
M. ovalifolia Dinter & A. Berger	Namibia and S.W. Angola
M. stenopetala (Baker f.) Cufod	Kenya and Ethiopia
Shrubs and Herbs With Tubers	
M. arborea Verdc.	North Eastern Kenya
M. borziana Mattei	Somalia and Kenya
M. longituba Engl.	Somalia, Kenya, Ethiopia,
M. pygmaea Verdc	North Somalia
M. rivae Chiov.	Ethiopia, Kenya
M. ruspoliana Engl.	Ethiopia, Somalia, Kenya

nutritious factors (https://www.fao.org/traditional-crops/moringa/en/). A successful mission entitled "Million Moringa Project: Moringa as a Viable Solution against Poverty, Malnutrition and Deforestation" was started in the Philippines by the UN to establish one million Moringa Trees that promote reforestation besides supporting business and revenues.

Further, this mission also focused on distributing Ready-to-Eat Therapeutic Food to the needy people living in the underdeveloped regions of the Philippines (https://sustainabledevelopment.un.org/partnership/?p=35666). Therefore, in addition to its use as a nutritionally and medicinally important food supplement, Moringa cultivation can be also used to ensure the livelihood of people living in developing countries.

Several philanthropists and non-government organizations (NGOs) around the world (including Trees for Life based at United States of America (USA), ECHO (Florida, USA), Church World Service (Indiana, USA), GIANT (Georgia, USA), Helen Keller International (Guinea), and Santé et Nature (Congo)) have promoted the nutritional benefits of Moringa and active in publicizing the use of *M. oleifera* in daily food basket as "Natural nutrition for the tropics".

It is also highlighted by Devkota and Bhusal (2020) that planting trees (particularly, Moringa) in large scale can sequester more carbon, and thereby such trees can mitigate the ill effects of climate change such as famine. Therefore, if Moringa is grown on a much larger scale, there is great potential to not only help to prevent the problems of changing climatic conditions, but also to secure the income of agriculturalists living in resource-limited environments.

1.2 Nutritional and Medicinal Values of Moringa

It is generally believed that medicinal plants provide a limitless and low-cost source of useful drugs to treat large numbers of life-threatening diseases. In the health and medical industry, phytochemicals are always preferred to cure diseases rather than synthetic molecules as phytochemicals have a physiological role in humans and are judged to be more compatible and safer for human consumption.

Traditional medicine has long been used as a source of health care and in fact, in developing countries, nearly 80% of the population still uses traditional plant products as medicine. In addition, though newer medicinal molecules are widespread, more than one-third of the global population does not have access to these drugs due to their high cost and non-availability in developing countries. Under this situation, it would be particularly desirable to develop plant-based products as a parallel system of medicine Aschemann-Witzel et al. (2020).

Besides using Moringa as nutrient-rich food supplements (see Chap. 3), all the parts of the Moringa plant have also been utilized as medicinal components over a long period of time (see Chap. 4).

Every part of *M. oleifera* including seeds, leaves, roots, flowers, stem, pods are edible (Table 1.3) and shown to have traditional medicinal value. A large array of benefits can be obtained from a single Moringa tree: leaf can be used as a source of nutrition for human and livestock, seed-oil are useful for cooking and biodiesel, seedcake (leftover after oil extraction) is generally used in water purification and as fertilizer and wood to build a shelter where there is no alternative wood is available.

Several historical reports also state that Moringa has long been advocated to treat anemia, diabetes, arthritis, asthma, typhoid fever, malaria and several skin diseases due to their richness of various phytoconstituents such as carbohydrates, glycosides, glucosinolates, flavonoids, carotenoids, isothiocyanates, polyphenol, saponins and alkaloids (Meireles et al. 2020).

Historical records depicted that the use of Moringa was reported as early as 150 BC and it was greatly cherished as a healthy food supplement in the ancient world. In order to sustain mental alertness and healthy skin, it has been evidently shown that the primordial kings and queens preferred Moringa leaves and fruit in their regular diet.

An extensive survey on the online literature database clearly indicated that the first report on the uses of Moringa was cited by Dalzell (1866). There are also other reports which documented that Moringa was originally cultivated by the Dravidians in South India and later it was grown

Table 1.3 Benefits of Moringa in Human Health

Parts of the Moringa	Description	Health Benefits	Documented evidence that supports its benefits
Leaves	Individual leaflets are 1–3 cm long with a blunt point or round at tip and short-point at the base. They are produced as compound leaves which are bipinnate or tripinnate or imparipinnate	Leaf powder or extract is used as a nutrient supplement to the human staple food besides fodder and feed supplement to livestock animals. Possesses cleansing ability and poultice the wounds and headaches. Also used to cure piles, fevers, scurvy, catarrh, sore throat, bronchitis, eye and ear infections. Moringa leaf extract is nowadays administrated to control glucose fluctuations and reduce glandular inflammation	Sha and Oza (2020)
Bark or Whole Stem	The matured stem is softwood and weighs light. Stem has whitish-gray, thick, soft, fissured and corky bark and exudes gums when it is injured	Extract or crude juice obtained from stem bark is used to treat eye maladies, skin redness, blistering, enlarged spleen, tuberculous glands found in the neck, tumors, ulcers and delirious patients. Juice extracted from the root bark is used to treat earaches and tooth pain due to tooth cavities	Brilhante et al. (2017)
Gums	Gum is produced by the Moringa stem when it is injured. Initially, gum is white in colour but later, when continuously exposed to sunlight, it turned into reddish brown or brownish black	Gum is frequently mixed with medicinally important phytochemicals (such as sesame oil) and used to treat tooth decay, relieve headaches, fevers, intestinal complaints, dysentery and asthma and treat syphilis and rheumatism	Singh and Kumar (2018)
Roots	Moringa has lengthy taproot with few thin lateral roots and has a unique pungent odour. On the whole, the root structures look like tuberous	Used primarily as antifertility, anti-inflammatory and antilithic agents; also act as a cardiac/circulatory tonic. Besides used as a laxative, abortifacient, treating rheumatism, articular pains, lower back or kidney pain and constipation and applied as a rubefacient, vesicant, carminative and stimulant in paralytic afflictions	Ezeamuzie et al. (1996)
Flowers	Fragrant, bisexual and yellowish white flowers borne on thin stalks as axillary clusters. Individual flower is 0.5 to 1 cm long and ~2 cm wide with five petals (unequal, dull white, sparsely veined, rounded), five stamens (staminodes) and a carpel (possessing unilocular ovary and slender style)	Have long been found in phyto-medical history. Mainly employed as an abortifacient, stimulant, aphrodisiac and cholagogue besides (1) to treat inflammations, muscle ailments, tumors, hysteria and enlargement of the spleen (2) to lower the serum cholesterol, phospholipid and triglycerides and (3) to reduce the lipid profile of liver, heart and vein and increase the excretion of fecal cholesterol	Siddhuraju and Becker (2003)
Pods	Moringa fruits are known as pods (usually pendulous with 30–120 cm	Immature and mature pods are regularly consumed as a vegetable and	Vieira et al. (2010)

(continued)

Table 1.3 (continued)

Parts of the Moringa	Description	Health Benefits	Documented evidence that supports its benefits
	long (up to 1 m length fruits are found in some variety, PKM2 (see Chap. 5) and ~1.8 cm wide fruit). Though fruits are found throughout the year, they are abundantly produced during Summer (April-May) in India. Initially, the pods are green in colour and rigid; however, upon maturity they become brown and opened longitudinally along their three angles and freeing the seeds from its inner compartments.	used in several culinary preparations with various spices and condiments. Pods possess all the nutrients found in leaves. Kids are usually preferred to have Moringa pods (referred to as drumsticks) and hence they have a unique place in customary medicinal preparations	
Seeds	Seeds are usually round to oblong in shape with 1 cm diameter and weighs about 0.3 gm. It holds brownish polymeric semi-permeable seed coat with three papery wings. Singly Moringa tree can produce 15,000 to 25, 000 seeds per annum and approximately 70% of them germinate within 14 days	Seeds are used for oil extraction (termed as ben oil) which is used in cosmetics and perfume industries besides as biofuels. Seed extract is used to decrease liver lipid peroxides. Antihypertensive compounds such as thiocarbamate and isothiocyanate glycosides were found in ethanolic extract of Moringa pods	Fahey (2005)

by the Indo-Iranian and Arians in every home. Interestingly, at the war front, Maurian warriors of India were fed with Moringa leaf extract to increase their fighting efficiency as this Moringa tonic was believed to provide extra vigour and get rid of them from trauma (Gopalan et al. 1971).

To support the preceding lines, recently, Shah and Oza (2020) provided literature evidence at the molecular level and concluded that extracts/fractions/phytoconstituents isolated from various parts of *M. oleifera* can be used to cure wound healing, inflammation, atherosclerosis, cancer, diabetes, arthritis and atopic dermatitis.

Under the prevailing viral pandemic conditions, the concept of immunity boosting is now gaining widespread attention. As the mass devastation of COVID-19 is taking a toll on the human population, a healthy immune system is a major reason for the survivors. Hence, there is a huge surge in immunity-boosting foods.

Moringa is regarded as one of the best foods for not only building immunity but also preventing and curing nearly 300 diseases (Prasad and Ganguly 2012). Evidence collected by Mehwish et al. (2020) confirmed the importance of Moringa as a functional food supplement and highlighted its immunomodulatory potential for human health. As a matter of fact, Moringa powder are generally used as a food fortification agent owing to its rich protein and high content of dietary antioxidants, ascorbic acid, β-carotene, α-tocopherol and mineral elements such as calcium, iron and potassium (more detailed Moringa's nutrient description is provided in Chap. 3).

Therefore, Moringa is emphasized as a superfood in the recent literature. Though the word has not yet been officially classified, superfoods are said to encompass rich sources of nutrients and bioactive compounds (including vitamins, minerals and antioxidants) that ensure all kinds of positive characteristics to mankind. The other examples of superfoods are hemp and chia seeds, cacao beans, goji berries, bee pollen, wheatgrass, coconut oil, seaweed and algae extracts.

In contrast, Dutch Food Centre (www.voedingscentrum.nl), the leading authority that offers scientific evidence for healthy, safe and more sustainable food choices, highlighted that the term superfood is rather a misleading word, as there is no sole food that can provide all the essential nutrients for mankind. At this junction, it is worth mentioning that owing to the richness of essential nutrients and other health benefits, Moringa can be termed as a superfood.

Finding a food crop that keeps man healthy and active has been the foremost quest since the beginning of human civilization and it still actively continues even in this modern era of research and technology. To this end, Moringa has been noticed as a promising tree as it has successfully cleared all the tests of nutritional benefits, bioavailability and medicinal properties and environmental safety and hence reputedly referred to as mother's best friend and miracle tree.

1.3 Promising Moringa Cultivars and Their Production

Globally, India has the largest area under Moringa (~38,000 ha) and has an annual production of 2.2 million tonnes of tender fruits (Pandey et al. 2011). In India, the state of Tamil Nadu possesses broad genetic diversity (in both annual and perennial types) and diversified geographical areas for Moringa cultivation and it is the pioneering highest Moringa producing state. A large array of diversified Moringa ecotypes or accessions (to name a few: Yazhpanam (Sri Lankan) Moringa, Chavakacheri Moringa, Nattu Moringa, Karumbu Moringa, Semmoringa and Kodikaal Moringa) are being cultivated commercially in Tamil Nadu, for its immature pods. Apart from the perennial types, two short duration annual varieties, viz. PKM1 and PKM2 have been released from Tamil Nadu Agricultural University (TNAU), Coimbatore, India, for commercial cultivation and are globally popular (see Chap. 5 for more details).

The cultivation details of moringa for the leaf as well pod production that are being followed in Tamil Nadu, India were explained by Balakumbohan et al. (2020). Usually, annual Moringa is propagated through seeds and seed rate of 4 kg seed/hectare is required for leaf production. Alternatively, moringa seedlings produced in polybag nursery or pro tray nursery can also be transplanted. Though Moringa is a hardy plant, it responds well under irrigated conditions. Especially in the case of leaf production, maintenance of optimum soil moisture enhances the speedy biomass development and improved green leaf yield.

1.4 Need for This Book

One of nature's gifts to men- and women- folk is *Moringa oleifera* as it is the most nutrient-rich and multipurpose plant among the food crops. The demand for this "green superfood" has immensely increased during the recent past owing to the health awareness that spawned among the public due to the COVID-19 pandemic. Scopus® (www.scopus.com), which is considered to be the largest abstract and citation database of peer-reviewed literature, was searched on 03 December 2020, for all citations with the keyword "moringa". The results of the Scopus search revealed 4,933 results for the period 1866–2020. On the other hand, searching Google Scholar (another web-based tool that is generally used to broadly search for scholarly literature) has resulted in 66,100 results for the same keyword, "moringa", on the same day. According to Scopus, Moringa research has experienced a surge of publications: eighty percent of all published materials since 1866 has occurred during the past 10 years (2009–2018), and over half of that is published recently, i.e. during 2014–2018 (Mainenti 2018).

This implies that a comprehensive and concise collection of all the scientific information on Moringa in one single book would be of great value to the Moringa researchers.

A wide array of genetic variations in Moringa for their local adaptation has been documented during its spread from the Indian subcontinent to throughout the tropical and subtropical world. However, rigorous and localized studies are yet to be conducted for the enhanced leaf and pod production and to test the leaf's or pod's nutritional content and their bioavailability. Similarly, controlled experiments must be performed to document the bioavailability and effectiveness of nutrients present in Moringa leaves to mankind. In addition, *M. oleifera* has also been shown as a promising source for second-generation biodiesel. Thus, from the biodiversity, ethnobotanical, dietary, pharmacological and socio-economic perspectives, conservation of the Moringa is of immense concern. Consequently, there is a need to bring all the above perspectives and information in one single book.

Likewise, in order to ensure the nutritional and other properties (such as oil content) in the Moringa cultivars that are evolved for the given environment, it is imperative to incorporate the current genomics tool in the Moringa genetic improvement program. Such a molecular breeding approach, indeed, increase the efficiency of the breeding strategy that focuses on the evolving novel Moringa cultivars with improved nutritional properties.

Therefore, the ultimate aim of this book is to provide updated and all-inclusive data and evidence for Moringa botany, cytogenetical analysis, genetic resources and diversity, classical genetics, traditional breeding, tissue culture, genetic transformation, whole-genome sequencing, comparative genomics and elucidation on applications of functional genomics, nanotechnology, bioinformatics, processing and value addition besides providing perspectives of medicinal and therapeutic properties of Moringa.

This book will be useful to both the academia and private companies interested in horticulture, molecular breeding, pathology, entomology, physiology, in vitro culture, genetic engineering and functional genomics in Moringa and other vegetable crops. This book will also be useful to entrepreneurs and business people belonging to the seed and pharmaceutical industries.

Ensuring food and nutritional security in addition to safeguarding human health in the under-developed rural area is recognized as an enormous and complex huge task. To achieve this goal, a collaborative effort rendered by the private sector, smallholders, associations, larger companies and public institutions is essential. Initiatives (such as a comprehensive scheme that provide knowledge and resources to the rural farmers) must be originated from the authorities of all the developing countries to improve the livelihood and standards of rural life. They should organize routine workshops (such as the series of workshops organized by TNAU under EDII- Periyakulam Horti-Business Incubation Forum (https://ediiphbif.org/)), train them in Moringa harvesting and processing and pay them according to their product quality and quantity. This type of empowerment will ensure economic security besides nutritional security among the family members. It is sincerely expected that this book will be an initiative towards this goal.

References

Alexandratos N, Bruinsma J (2012) World agriculture towards 2030/2050: the 2012 revision. ESA Working paper FAO, Rome

Aschemann-Witzel J, Gantriis RF, Fraga P, Perez-Cueto FJ (2020) Plant-based food and protein trend from a business perspective: markets, consumers, and the challenges and opportunities in the future. Crit Rev Food Sci Nutr, 1–10

Balakumbahan R, Sathiyamurthy VA, Janavi GJ (2020) Moringa leaf–a super food. Biotica Res Today 2 (6):438–440

Brilhante RSN, Sales JA, Pereira VS, Castelo DDSCM, de Aguiar Cordeiro R, de Souza Sampaio CM, …& Rocha MFG (2017) Research advances on the multiple uses of Moringa oleifera: a sustainable alternative for socially neglected population. Asian Pacific J Trop Med 10(7):621–630

Dalzell NA (1866) Remarks on the genus Moringa. Bot J Linn Soc 9(37):302–304

Devkota S, Bhusal KK (2020) *Moringa oleifera*: a miracle multipurpose tree for agroforestry and climate change mitigation from the Himalayas–a review. Cogent Food Agric 6(1):1805951

Ezeamuzie IC, Ambakederemo AW, Shode FO, Ekwebelem SC (1996) Antiinflammatory effects of Moringa oleifera root extract. Int J Pharmacognosy 34(3):207–212

Fahey JW (2005) Moringa oleifera: a review of the medical evidence for its nutritional, therapeutic, and prophylactic properties. Part 1. Trees for life J 1(5):1–15

Gopalan C, Rama Sastri BV, Balasubramanian SC (1971) Nutritive value of Indian foods, ICMR, Hyderabad, India, (National Institute of Nutrition), p 56

Gruda N, Bisbis M, Tanny J (2019) Impacts of protected vegetable cultivation on climate change and adaptation strategies for cleaner production–a review. J Clean Prod 225:324–339

Mainenti DAVID (2018) Moringa oleifera: prolific research of a superfood. Information Science, Bibliometrics. Retrieved 4(26)

Mehwish HM, Riaz Rajoka MS, Xiong Y, Zheng K, Xiao H, Anjin T, He Z (2020) Moringa oleifera–a functional food and its potential immunomodulatory effects. Food Rev Int, 1–20

Meireles D, Gomes J, Lopes L, Hinzmann M, Machado J (2020) A review of properties, nutritional and pharmaceutical applications of Moringa oleifera: integrative approach on conventional and traditional Asian medicine. Adv Tradit Med, 1–21

Palada MC (1996) Moringa (Moringa oleifera Lam.): a versatile tree crop with horticultural potential in the subtropical United States. HortScience 31(5):794–797

Pandey A, Pradheep K, Gupta R, Nayar ER, Bhandari DC (2011) 'Drumstick tree' (*Moringa oleifera* Lam.): a multipurpose potential species in India. Genetic Resour Crop Evol 58(3):453–460

Prasad A, Ganguly S (2012) Promising medicinal role of Moringa oleifera: a review. J Immunology Immunopathol 14(1):1–5

Ray DK, Mueller ND, West PC, Foley JA (2013) Yield trends are insufficient to double global crop production by 2050. PLoS ONE 8(6):

Rouhani S, Griffin NW, Yori PP, Gehrig JL, Olortegui MP, Salas MS,… & Kosek MN (2020) Diarrhea as a potential cause and consequence of reduced gut microbial diversity among undernourished children in Peru. Clin Infectious Diseases 71 (4):989–999

Shah KH, Oza MJ (2020) Comprehensive review of bioactive and molecular aspects of Moringa Oleifera lam. Food Rev Int, 1–34

Siddhuraju P, Becker K (2003) Antioxidant properties of various solvent extracts of total phenolic constituents from three different agro-climatic origins of drumstick tree (*Moringa oleifera* Lam.). J Agric Food Chem 15:2144–2155

Singh B, Kumar A (2018) Hydrogel formation by radiation induced crosslinked copolymerization of acrylamide onto moringa gum for use in drug delivery applications. Carbohyd Polym 200:262–270

Vieira AMS, Vieira MF, Silva GF, Araújo ÁA, Fagundes-Klen MR, Veit MT, Bergamasco R (2010) Use of *Moringa oleifera* seed as a natural adsorbent for wastewater treatment. Water Air Soil Pollut 206(1–4):273–281

Botanical Descriptions of *Moringa* spp.,

2

N. Manikanda Boopathi and Bashir Yusuf Abubakar

Abstract

The *Moringa* genus of the Moringaceae family is monogeneric of the Dicotyledaneae and it contains 13 species. *Moringa* spp., are geographically distributed in several tropical countries. For example, *M. drouhardii* and *M. hildebrandtii* are found in Madagascar, *M. stenopetala* is noticed in Kenya and Ethiopia, *M. ovalifolia* is located in Namibia and Angola, *M. concanensis* is largely reported in India, Pakistan and Bangladesh, *M. oleifera* is found in India, *M. peregrina* is observed in Arabia and Red sea area. On the other hand, *M. arborea, M. longituba, M. rivae, M. pygmea, M. borziana* and *M. ruspoliana* are indigenous to Kenya, Somalia and Ethiopia. Each Moringa species has unique morphological features and this chapter describes the botanical descriptions of all *Moringa* spp., in detail and provides useful traits that can be used to distinguish each *Moringa* species.

N. M. Boopathi (✉)
Department of Plant Biotechnology, Centre for Plant Molecular Biology and Biotechnology, Tamil Nadu Agricultural University, Coimbatore, India
e-mail: nmboopathi@tnau.ac.in

B. Y. Abubakar
Department of Botany, Ahmadu Bello University, Zaria, Nigeria

2.1 Introduction

In the literature, "horseradish" tree, "drumstick" tree, "ben-oil tree", "benzoil tree", "cabbage tree", "mother's best friend" and "miracle tree" are synonymously referred to Moringa (botanically, *M. oleifera*). Though it was originated in the Indian subcontinent, now Moringa is cultivated elsewhere in the world (especially in countries of tropics and sub-tropics) owing to its numerous utilities. It is an affordable and a rich source of proteins, multi-vitamins (more importantly vitamin A, B and C), calcium, potassium and possesses an exclusive blend of quercetin, kaempferol, zeatin, sitosterol, caffeoylquinic acid and other medicinally imperative secondary metabolites. These nutrients and medicinal compounds are found in each part of this tree and hence they are considered as an inexpensive reservoir of dietary supplements.

Owing to these properties, Moringa has been found in widespread applications ranging from antioxidant, anti-inflammatory, antidiabetic, antibacterial, antifungal, antiulcer antihypertensive, antipyretic, antiepileptic, antitumour, anticancer, antispasmodic to cardiac and circulatory stimulant with diuretic, hypocholestemic and hepatoprotective properties. Besides its medicinal benefits, Moringa is also employed in reforestation, fodder and/or feed, water purification and biosorbent (to sequester heavy metals from polluted water). Ben oil extracted from Moringa

N. M. Boopathi et al. (eds.), *The Moringa Genome*, Compendium of Plant Genomes,
https://doi.org/10.1007/978-3-030-80956-0_2

seeds has huge market demand, as it is often used in lubricants, skin lotions and more notably in the production of biodiesel.

Considering the aforesaid benefits and market value of Moringa, it is imperative to understand the botany and charactcristics of Moringa germplasm for its further genetic improvement of various economically important traits.

2.2 Classification and Evolution

Moringa is the only genus (monogeneric) under the family Moringaceae. However, among the angiosperms, Moringaceae is considered as the greatest phenotypically varied family (Olson 2001). For example, though there are only 13 species of *Moringa*, found in the entire dry tropics of the old world, they possess a vast range of life forms or habits: (i) huge bottle-shaped trees in Madagascar and Africa, (ii) slender trees in Arabia and India (iii) shrublets with ephemeral shoots and (iv) underground tubers in northeast Africa (Olson and Carlquist 2001).

The plant is of the Hindustan centre of crop origin, indigenous to the western and sub-Himalayan tracts of India and Pakistan (Ramachandran et al. 1980). However, the species are geographically distributed in tropical countries: *M. drouhardii* and *M. hildebrandtii* are found in Madagascar, *M. stenopetala* is noticed in Kenya and Ethiopia, *M. ovalifolia* is located Namibia and Angola, *M. concanensis* is largely reported in India, Pakistan and Bangladesh, *M. oleifera* is found in India, *M. peregrina* is observed in Arabia and Red sea area. On the other hand, *M. arborea, M. longituba, M. rivae, M. pygmea, M. borziana,* and *M. ruspoliana* were believed to be originated in Kenya, Ethiopia and Somalia.

By employing molecular phylogenetic data, a sister family relationship of the Moringaceae with Caricaceae was proposed during the dawn of the twenty-first century based on wood anatomy data which presented phylogenetic information at an interspecific level (Olson 2001). Later, a comprehensive data set on three genes, viz. *18SrDNA* (1855 bp), *rbcL* (1428 bp) and *atpB* (1450 bp) representing a total of 4733 bp were inferred and confirmed the closeness of *Moringa* and *Carica* genera (Soltis et al. 2000). These two genera formed a subcluster at 94% similarity index of a jackknife consensus tree.

Further, a sister taxon relationship between Caricaceae and Moringaceae has also been reported based on both morphological and molecular characters (Olson 2002). This study has indisputably established that obvious stalk glands are found at the points where (i) the petiole intersects the stem, (ii) pinna intersects the rachis and ((iii) leaflet intersects an axis in *Moringa* species and such structures were also noticed in Caricaceae.

As members of these families produce muster oil, in a broader sense, Caricaceae and Moringaceae should be placed within the cluster of Brassicacaceae (Soltis et al. 2000). DNA-sequence-based analysis has provided strong support to this close relationship and other mustard oil producing families such as Caper family (capparaceae) and Nasturtium family (Tropaeolaceae) have also been found to have such relationship (Olson 2001).

Actually, in earlier reports, the monogeneric family of Moringaceae has been grouped in or near Capparales because of the shared presence of glucosinolates and mustard oils and other characters such as the presence of zygomorphic flower, gynophores, parietal placentation and lack of endosperm (Decraene et al. 1998). Based on morphological details of the plants, no one had ever suspected that the families of Moringaceae and Caricaceae (papaya family) were so closely related. After publishing the anatomical experiments on other characters, it was strongly established that they share a strong similarity in their wood anatomy.

2.3 Botanical Descriptions of *M. oleifera*

The first description of *Moringa oleifera* was reported in 1785 by the French naturalist Jean-Baptiste Lamarck. It was believed that the genus name "Moringa" might be derived from a Tamil word "murungai", which means that "twisted hand length structure of the young Moringa fruit". In Latin, the word, "oleum" refers "oil" and "ferre" means "to bear".

Morphological and comparative genomics studies have indicated that *M. oleifera* was most closely related to *M. concanensis* and both of these species might be diverged from the common ancestor of *M. peregrina* (Olson 2002).

As early as 1785, Swedish Biologist, Linneus thought that the *M. oleifera* tree (Fig. 2.1) was a member of the legume family. This was based on the fact that bilaterally symmetrical flowers resembled legume flowers, superficially (Olson and Carlquist 2001). Another commonality between these two groups is the production of pinnate leaves. This view was accepted by some prominent botanists until the mid-1800s, while others advocated placement near the violet family. Olson (2002) reported that a few of botanists declared *Moringa oliefera* as a member of the bignonia family based on its long woody pods and wing seeds. On the other hand, Hutchinson in 1926 has included *Moringa* in Capparidales. Later, in 1964, Engler concurred with Bassey's report that was published in 1915 and placed *Moringa oliefera* under the suborder Moringaceae and order Rheoadales (Ramachandran et al. 1980).

Generally, *Moringa oleifera* can grow up to 8 m and thus has been described as a shrub or small tree. It usually has a low branching habit with a spreading crown and can be effortlessly identified by its tripinnate leaves and lengthy fruit (up to 60 cm; exceptionally one variety, viz. PKM2 can grow up to 1 m) with a triangular cross-section. The bark of the plant is smooth, grey to brownish, roughly lenticellate, with slash green outside and yellowish underneath. The trunk exudes, when injured, a white opaque gum turning dark red or brownish red when exposed to the air and sun. Stems are more or less densely pubescent to puberuluos, lenticellate and grey.

M. oleifera produces composite leaves that are alternate, bipinnate or partly tripinnate (which measures 30–70 cm long) with two to six pairs of opposite pinnate bearing leaflets in three to five pairs. Leaflet at the terminal end is larger than the other leaves, may or may not be glaborous with ovate to elliptic size. The leaves are one to two cm long and 0.5–1.5 cm in width with a rounded apex and base.

However, wide variation in morpho-anatomical features of the *M. oleifera* accessions has been noticed in Asia and northern Nigeria (for example, the accession 16BAU possessed 60.3 cm leaf length with an epidermal cell width of 32.6 μm on the adaxial surface and has pubescent petiole with a length of 12–15 cm) and it was suggested that such genetic variation can be explored to genetically improving the Moringa cultivars (Abubakar et al. 2011).

Inflorescence of *M. oleifera* is a terminal raceme, borne at the base of a leaf and scattered at the ends of branches and 10–15 cm long. Flowers are cream (sometimes red tingled), pedicellate (3–5 mm long), with 5 sepals, 5 unequal petals (the two upper ones being

Fig. 2.1 Morphological appearance of a typical *M. oleifera* tree growing in **a** India and **b** northern Nigeria

Fig. 2.2 Morphological features of *M. oliefera* flower: **a** inflorescence and **b** close look of opened flower

smallest) bent or curved, slightly longer than the sepals (Fig. 2.2).

Usually, medium-sized (1.0–1.5 cm long and 2.0 cm broad), fragrant and hermaphroditic flowers are found in *M. oliefera*. Flowers are produced as spreading or drooping clusters on slender, hairy stalks that have a length of 10 to 25 cm. It has five yellowish white, unequal, sparsely veined petals.

Flowering begins six months after planting. When there is a uniform seasonal temperatures and rainfall pattern, peak flowering occurs twice a year; in certain pockets, flowers are produced round the year. However, under a cooler climate, flowering occurs once a year (mostly during April–June).

Fruits of *M. oleifera* are elongated, linear, with 3–4 ribs, 2 furrows on each side, 30–50 cm long and 2 cm across, brown or beige to grayish when completely matured (Fig. 2.3). Seeds are spherical, black (sometimes brown), 7–8 mm in diameter with 3 to 4 papery wings that are yellowish grey.

M. oliefera produces hanging fruit that has three-sided brownish-green capsule of 20–50 cm size (at maturity) which contains dark brown and globular seeds with $\sim$ 1 cm diameter. The seeds have three whitish papery wings and are dispersed by wind and water (Fig. 2.3c).

Ben oil is an edible oil extracted from mature *M. oliefera* seeds and approximately 38–40% oil can be extracted from the seeds. Ben oil is clear and odorless after refining and can withstand rancidity for a longer time compared with other vegetable oils. In the *M. oliefera* seed cake, residues obtained after the extraction of oil are frequently applied as a fertilizer or used as a flocculent to purify water. Moringa seed oil is also considered a potential biofuel.

M. oleifera grows well in India and Nigeria and other tropical countries as an acclimatized crop at lower altitudes (Bosch 2004). Though it has been reported in East Africa at 1350 m altitude, it has also been found in 2000 m in Zimbabwe. Compared to other crops, *M. oliefera* is invariably recognized as drought tolerant and it can thrive well with as little as 500 mm annual rainfall and it has also been found to be tolerant (to some extent) to frost. Though *M. oleifera* is cultivated in a wide range of soils, preferably fertile and well-drained soils are most suitable.

2.4 Botanical Descriptions of *M. stenopetala*

Moringa stenopetala is a deciduous tree indigenous to Ethiopia and Kenya and it is known in the name of African Moringa or cabbage tree (Leone et al. 2015). It has peculiar characteristics such as bottle-shaped trunk, long twisted seed pods and edible leaves resemble cabbage (and hence it got the name cabbage tree) and found to be resistant to water stress. The majority of the wild *M. stenopetala* has been uprooted in Ethiopia; however, in the terraces of the

Fig. 2.3 Morphological feature of fruit or pods of *M. oleifera* **a** full-grown pods **b** completely matured pods and **c** seeds

Ethiopian Highlands, mainly in the Konso region, few people grow *M. stenopetala* as a crop.

Edmund G. Baker has first described this species as *Donaldsonia stenopetala* in 1896 based on the type of specimen collected by Donaldson Smith from the northeastern shore of Lake Turkana. A similar kind of specimen collected from Kenya has been named as *Moringa streptocarpa* by Emilio Chiovenda. Both names have been officially renamed as *Moringa stenopetala* in 1957 by Georg Cufodontis since they were morphologically similar. The species name *stenopetala* was derived from the Greek words, *stenos* (which means narrow or straight) and *petalum* (which denotes petal) to illustrate the shape of its petals (Habtemariam 2017).

M. stenopetala relatively lives a longer period of time (nearly or more than 100 years) and continuously produces leaves (and/or reproductive) throughout its life. It can quickly grow and reach a height of 3 m in its first year and can continue to grow up to 12 m and in some exceptional cases to 15 m. It attains reproductive maturity within two and half years and continues to flower and set fruit throughout its lifetime. Owing to the sweet fragrance of the flowers, birds and insects are attracted and it provides a great opportunity for cross-pollination.

Seeds are generally dispersed through either wind or water. The tree is an important source of food stuff for the animals living in the surrounding regions: foliage is preferred by the wild and domesticated ruminants as their staple food and fragrant flowers are the favorite source of nectar for honeybees.

Its perennial tree trunk is bloated at the bottom and normally branched into fork-like structure with a diameter of 1 m, which leads to Caudiciform or bottle-shaped structure (Fig. 2.4). Smooth and whitish to light gray or silver bark is found in the stem which has softwood at bottom. *M. stenopetala* possesses younger shoots with a characteristic dense leaf with velvety pubescence and it's crown is sprawling and heavily branched.

Generally, *M. stenopetala* produces leaves that are light green in colour that are ~5.5 cm long and attached alternatively to the stem by short petioles. Leaves are either bi- or tripinnate in composition that has five pairs of pinnae and each pinna has three or a maximum of nine leaflets. The individual leaflet is 3.5–6.5 cm long and 2–3.5 cm width with a shape of elliptical to ovate that has acute tip and round-to-cuneate bottom. Another striking character is the presence of stipule-like extrafloral nectaries at the base of the leaflet.

A thick and fragrant inflorescence arranged as a dense panicle (with a length of 60 cm) is found in *M. stenopetala*. Bisexual, radially symmetrical, and pentamerous flowers are found in the

Fig. 2.4 Morphological appearance of a typical *M. stenopetala*

panicle. The calyx is cream or pink colored, polysepalous with 4–7 mm long sepals. White or pale-yellow or yellow-green and polypetalous corolla is found with 8–10 mm long petals that are nearly oblong in shape.

There are five stamens in each flower with 4–6.5 mm long white filaments attached with 2 mm long yellow anthers and several variably sized shorter staminodes. *M. stenopetala* has a densely haired and superior ovary which is 2 mm long and ovoid in shape which extended into a silky cylindrical style with a stigmatic lobe (Verdcourt 1986).

M. stenopetala fruits are elongated pods (19.7–50 cm long with 1.8–4 cm width) which are twisted and bright green during early development but gradually become straight (however, in certain cases it forms torose (bulbous) around the encapsulated seeds). When pods are matured, they become reddish with discolored dots. Upon maturity, pods are dehiscent and split open along three prominent grooves and each pod can release up to 20 seeds.

The seeds are oblong to triangular which is overcoated with cream (or brownish) lined with husk with three prominent papery wings. Seeds are usually 2.5–3.5 cm long with 1.5–2 cm breadth and have whitish grey coloured smooth kernel.

2.5 Botanical Descriptions of *M. concanensis*

Moringa concanensis Nimmo., also belongs to the family Moringaceae, is widely found throughout the states of Rajasthan and Tamil Nadu, India.

M. concanensis is a small tree (but can grow 6–12 m in height) with branches that hang almost near to the ground (Fig. 2.5). It has a circular stem that produces few erect and small branches with yellow wood. Branch produces two pinnate leaves with 5–8 pair opposite leaflets and each leaflet is broad, ovate to elliptic having 1–2 cm width and 1–3 cm length. It has a taproot system.

Fig. 2.5 Morphological feature of *M. concanensis*

The inflorescence of *M. concanensis* is a panicle that bears flowers with the following characteristics: petals are white in colour with purple streaks, oblong to obovate (1.5 × 0.5 cm) and unequal. It has five fertile stamens and declinant staminodes. *M. concanensis* has stipitate ovary and capsules beaked fruits with winged seeds.

2.6 Botanical Descriptions of *M. drouhardii*

Moringa drouhardii Jumelle., is a widespread species of southwest Madagascar and it is popularly identified as 'bottle tree' in those regions. *M. drouhardii* is mainly found in the Madagascar spiny thickets ecoregion (more precisely the limestone cliffs to the east of Lake Tsimanampetsotsa on the Mahafaly Plateau) in limited numbers (just in hundreds) that too with scattered patterns.

M. drouhardii is a deciduous tree that may have a height of 5–10 m (however, some of them were found grow up to 18 m). In order to store water, the stem use to swell, and the stem is usually unbranched (but has short branches at the top). The leaves are feathered on the stem (as that of ferns) which is white and boated.

The fast growing nature of the *M. drouhardii* helps it to rapidly occupy the open spaces in the forest and when cultivated it can grow at a rate of >1 m per year. Within three years, it can reach 3–4 m and after it starts to bear pods.

M. drouhardii is usually cultivated in local villages, especially around the traditional tombs. Owing to their significant medicinal benefits, people living in Atsimo-Andrefana Region (southwestern Madagascar), use *M. drouhardii* seeds (rich in edible oil and flocculating proteins) and leaves in their regular diet.

M. drouhardii is a deciduous tree that can attain a height of 18 m with a swollen bottom with short branches on the top of the tree. It has whitish bark that has rich resin. *M. drouhardii* has alternate leaves that are three pinnate. It does not contain stipules but has 10–15 cm long petiole, 2–3 cm long stalks of pinnae and 3–4 mm long petiolules. All petioles are glabrous with glands at the bottom. Leaflets are generated in opposite directions with bright green. Usually, the leaf is ovate to oblong and 15–30 × 5–12 mm size. Leaf base is cuneate with acute apex acute and glabrous.

Inflorescence of *M. drouhardii* is axillary, lax and 30 cm long panicle bears several numbers of flowers. Regular and bisexual flowers are found in 5-merous and yellowish white colour. Usually, the pedicel is glabrous, 2 mm long with free, obovate and 5–6 mm × 2 mm sized sepals that are narrowed at the base but have a round apex. Petals are free and 7–10 mm × ~2 mm sized ovate with incurved apex. Petals are glabrescent outside but inside it is slightly short-hairy. *M. drouhardii* has five free and hairy stamens that are 6–8 mm long. They are alternating with five staminodes which are ~4 mm in length. It has a superior, stalked and ~1.5 mm long single-celled ovoid with a slender style of 3–4 mm long.

M. drouhardii fruit is an elongated capsule that may have 30 to 50 cm length with an approximately trigonous structure which is narrowed between the seeds with a beak, glabrous, dehiscent with three valves. Seeds are glabrous and whitish with trigonous to ovoid that has the size of 2–2.5 cm × ~2 cm.

2.7 Botanical Descriptions of Other *Moringa* spp.

There is another tall Moringa tree in Madagascar with a massive, water-storing stem that can grow up to 20 metres height, which has been classified as *Moringa hildebrandtii* Engler. It closely looks like another well-known but unrelated baobab trees (Adansonia), which are also reported in Madagascar. The earliest specimen of *M. hildebrandtii* was collected by the Western botanist, Hildebrandt, in 1880 from the town of Trabonjy in northwestern Madagascar. Later, other botanists have documented *M. hildebrandtii* in villages of the west coast. Interestingly, *M. hildebrandtii* produces a metre-long pinnately compound leaves with a distinctive deep red stem tip and leaf rachis, when the plant is young. It generates tiny whitish flowers in large rachis.

Moringa ovalifolia Dinter ex Berger., is also having bloated trunks with white or dull white coloured stem. It is generally found alone on barren rocky or hilly regions and hence sometimes referred to as "ghost tree". *M. ovalifolia* is reported from central Namibia to southwestern Angola, especially on the rocky surfaces; however, in rare cases, it has been identified on soil (for example, at the Sproukieswoud (Fairy Forest) area of Etosha park. Similar to the other bottle shaped Moringa species, the leaves of *M. ovalifolia* are arranged in a palmately compound pattern and possess oppositely arranged three to five leaflets.

Arabian Peninsula Horn of Africa and Southern Sinai, Egypt are indigenous to another Moringa species, viz. *Moringa peregrina* (Forssk.) Fiori., and it has been reported widely in Arabia, Egypt and Syria. It prefers to grow on rocky valleys and cliffs in drier areas. It is a 6–10 m height deciduous tree with bigger leaves that are produced in thin pendulous branches. *M. peregrina* bears five-petaled flowers that are white or streaked red or pink. The pods of this species are distinctive which is usually up to 30 cm long, narrowly cylindrical with deep longitudinal grooves and bear on the tree throughout the year. Upon ripening they split into three valves and disperse the relatively large seeds that are white in colour (Dadamouny et al. 2016).

Allan Radcliffe-Smith and Peter Bally have documented yet another Moringa species, *Moringa arborea* Verdcourt., in 1972, which was grown as a single tree on a rocky canyon in northeastern Kenya near the Ethiopian border. The tree bears clusters of pale pink and wine-red flowers throughout the year and almost it covers the entire tree. The pods or young fruits look like a yard-long string bean. Kenyans and Ethiopians use the roots of this species (which are thick, fleshy and pungent-smelling) as a medicine.

Moringa rivae Chiovenda., is originally found in southern Lake Turkana to Mandera District in Kenya and all-over southeastern Ethiopia, however, not reported in deep interior Somalia. There are two distinctive subspecies: *rivae* possess flowers that are creamy petals with brownish tips and short fruits, whereas *longisiliqua* produce yellow flowers and relatively larger fruits. Usually, the plant seems to be shrubs (which can attain a height of 3 m) and has thicker and wider than usual roots which look like tuberous (or greatly swollen). However, certain subspecies of *longisiliqua* reported to be an arborescent habit and can grow up to 6 m.

Locally, names of the above Moringa species are used interchangeably, since *M. rivae* is similar to the features of *M. arborea* and *M. borziana*. Another striking fact is that the local name "lorsanjo" was used to refer to both *M. stenopetala* (which has a huge structure) and *M. rivae* (which is a smaller tree) by Samburu, Turkana and Rendille people of Kenya. In the same way, all the five Moringa species are denoted as "wamo" by Somalis of northeastern Kenya.

The coastal regions of southern Kenya to the region of Kisimayu in southern Somalia (especially within 100 miles of the coast), a novel Moringa species, viz. *Moringa borziana* Mattei., has been reported by Giovanni Ettore Mattei in 1908. It is a relatively little tree and has one or two stems that can attain a height of one meter and possess a large root tuber (which may reach 1–1.5 m depth in the soil). Interestingly, the stems are dried up (die back) after a couple of years (but not the tuber) and after some years it has the resurgence to grow into a tree. *M. borziana* flowers are greenish cream to yellow with brown smudges on the petal tips and scents unique sweet fragrance. When there is no flower or pod, *M. borziana* may be misclassified as *M. rivae;* however, the occurrence of *M. rivae* is never reported in the coastal shoreline.

Another Moringa species found in Somalia is *Moringa pygmaea* Verdc. It produces two to three pinnate obovate leaves with a size of 0.5 × 0.5 cm. *M. pygmaea* has flowers that are yellow or dull purplish brown and having cup-shaped receptacle. Their sepals are 0.9–1.2 cm long and the 1.2–1.3 cm long petals are usually with glabrous ovaries. The ciliate is at the apex. Though the flowers are similar to that of *M. borziana*, the upright fruits of *M. pygmaea* classified them as a unique species in

Moringaceae. All other species of Moringa produce hanging pendulant pods.

Moringa longituba Engler., is found in northeastern Kenya, southeastern Ethiopia and several parts of Somalia. It has a small shoot (60–75 cm) and a relatively large tuber. It generated pubescent 2–3 pinnate leaves with 2 pairs of pinnae and 3–5 leaflets per pinna. Leaves are generally oblong to obovate (2.5–6 × 1.5–4.5 cm), rounded to emarginate at the top, cuneate to subcordate at the bottom, glabrous or pubescent (when leaves are young). *M. longituba* flowers are bright red in colour, which appear when the plant is leafless and has a long, tubular hypanthium as the bases of petals and sepals are fused. The receptacle is tubular which is normally 1.3–3 cm long. It has 0.9–1.2 cm long sepals that are pubescent at the top and 1.2–1.5 cm long petals with a glabrous ovary. *M. longituba* produces pods with a size of 13.5–30 × 1.2–1.4 cm and has 1–1.3 cm long seeds with wings (measures about 2–3 cm). Like that of other Moringa species, *M. longituba* is also used as therapeutic plants in the Horn of Africa; especially, to cure intestinal disorders of camels and goats, *M. longituba* roots are internally administrated.

Yet another morphologically divergent species in Moringaceae family that are found in northern Somalia to southeastern Ethiopia (but rarely in northeastern Kenya) is *Moringa ruspoliana* Engler., which has simple pinnate leaves. Among the Moringaceae, leaflets of *M. ruspoliana* are the largest (can attain a diameter of 15 cm), thickest and toughest and also possess the largest flowers (reaching a length of 3 cm). During flowering, *M. ruspoliana* looks like *Erythrina* as their flowers are pink with green bases. Though *M. ruspoliana* develops thick taproot during the early period, later it swells and attains a large globular structure with four or few thick side roots. The matured *M. ruspoliana* can be described as a small tree (6 m length) with an octopus-like anchor system of long and fleshy tubular roots.

2.8 Conservation of *Moringa* spp., and Future Prospects

Of late, *Moringa* products are recognized worldwide as a precious herbal commodity. However, except *M. oliefera*, all the other 12 species are under endangered status and it is hard to find significant acreage of each species in the given region. This necessitates urgent actions on developing appropriate strategies and methodologies to conserve and sustainably utilize various *Moringa* species and place them as readily accessible and reliable reference germplasms.

To this end, all potential strategies, such as in situ and ex situ conservation and cultivation practices for effective resource management (i.e. good agricultural practices and sustainable utility plans) should be effectively employed.

Available datasets emphasized that most of the *Moringa* spp., are peculiar to the given regional environment which induces the synthesis of secondary metabolites that imparts specific medicinal property to the leaf, flower or root tubers. Such secondary metabolites may not be obtained using in vitro or plant tissue culture conditions, unless otherwise providing conducing expensive environments.

Therefore, in order to safeguard native Moringa germplasm with respect to their tight linkage of relationships with the natural world, in situ conservation of whole communities is essential. Furthermore, such efforts will increase the efficiency of preserving the target germplasm to its maximum diversity besides strengthening the association among conservations, utilizations and further explorations.

Globally, though, several ecosystem-oriented and species-oriented in situ conservation efforts have been focused on developing protected areas, very few efforts were employed in conserving the Moringa germplasm. Thus, there is a need to develop a sound conservation strategy by considering the rules, regulations, and potential compliance of Moringa germplasm within their growth environments.

It would also be desirable to employ biotechnical approaches in a routine breeding program (such as cell suspension culture, micropropagation, synthetic seed technology and molecular marker or genomics assisted breeding strategies; refer to other chapters of this book) to genetically enhance different *Moringa* spp., for yield and other nutritionally important attributes.

References

Abubakar BY, MuA'zu S, Khan AU, Adamu AK (2011). Morpho-anatomical variation in some accessions of *Moringa oleifera* Lam. from Northern Nigeria. African J Plant Sci 5(12):742–748

Bosch CH (2004) *Moringa oliefera* Lam. In: Grubben GJH, Denton OA (eds) Plant Resources of tropical Africa. Vegetables PROTA foundation, Wegeningen, Netherlands, Backhuys Publishers, Netherlands, pp 392–395

Dadamouny MA, Unterseher M, König P, Schnittler M (2016) Population performance of Moringa peregrina (Forssk.) Fiori (Moringaceae) at Sinai Peninsula, Egypt in the last decades: Consequences for its conservation. J Nat Conserv 34:65–74

Decraene LPR, Laet JD, Smets EF (1998) Floral development and anatomy of *Moringa oleifera* (Moringaceae): what is the evidence for a Capparlean or Sapindalean affinity? Ann Bot 82:273–284

Habtemariam S (2017) "1: Moringa stenopetala—botanical and ecological perspectives. The African and Arabian Moringa Species (1 ed.). Elsevier, pp 3–12

Leone A, Spada A, Battezzati A, Schiraldi A, Aristil J, Bertoli S (2015) Cultivation, genetic, ethnopharmacology, phytochemistry and pharmacology of Moringa oleifera leaves: An overview. Int J Mol Sci 16(12):12791–12835

Olson ME (2001) Introduction to the Moringa family. In: Fuglie LJ (ed) The miracle tree: multiple attributes of Moringa. CTA/CWS, Dakar, Senegal, pp 11–27

Olson ME (2002) Combing data from DNA sequences and morphology for a phylogeny of moringaceae (Brassicales). Syst Bot 27(1):55–73p

Olson ME, Carlquist S (2001) Stem and root anatomical correlations with life form, diversity, ecology and systematics in *Moringa* (moringaceae). Bot J Linn Soc 135:315–348

Ramachandran C, Peter KV, Gopalakrishnana PK (1980) Drumstick (*Moringa oleifera*): a multipurpose Indian vegetable. Econ Bot 34:276–283

Soltis DE, Soltis PS, Chase MK, Mort ME, Albach DC, Zanis M, Savolainen, Hahn WH, Hoot SB, Fay MF, Axtell M, Swensen SM, Prince LM, Kress WJ, Nixon KC, Farris JS (2000) Angiosperm phylogeny inferred from 18S rDNA, *rbcL* and *atpB* sequences. Botanical J Linnean Soc 133:381–461

Verdcourt B (1986) "Moringaceae". In Polhill RM (ed) Flora of Tropical East Africa. Flora of Tropical East Africa. CRC Press

Effect of Different Processing Methods on Nutritional Content of *Moringa Oleifera*

3

S. Amutha, K. Jothilakshmi, J. Devi Priya, and G. Hemalatha

Abstract

Moringa oleifera is widely considered as the tree of miracles since its leaves, roots, flowers, pods and seeds have been consumed traditionally as food and medicine. The leaves are inexpensive and abundantly available but are largely underutilized and often discarded. Several attempts have been made at this center in evaluating the effect of different processing methods of *Moringa oleifera* leaves and on formulating functional value-added foods that are Ready to Eat. Fresh Moringa leaves can be blanched, shade dried (to less than 10% moisture content), ground and then stored in airtight container. Such preparations of Moringa Leaf Powder (MLP) were examined for the final recovery of dried leaves, the impact of different drying procedures on final yield, proximate composition and mineral (Ca, Fe, K, P) content of the MLPs. Besides, several numbers of value-added foods that are Ready To Eat were formulated by adding the developed MLP into different food supplements. The nutrient content of dried Moringa leaves and powder after dehydration became a concentrated source of all the nutrients. Sensory analysis was carried with 50 semi-trained panelists, using a 9-point hedonic scale and all the products were well accepted. MLP retained appreciable amounts of nutrients (such as 17 g of protein, 3.8 g of crude fibre, 19622 μg of β carotene, 966 mg of calcium, 210 mg of phosphorus and 2.6 mg of iron content). It was also found that 100 g of MLP value-added foods provided 40–60 g of carbohydrate, 12.3–18.64 g protein and 1062–17,268 μg of β-carotene. Thus, it can be concluded that abundantly available and highly nutritious but largely underutilized *Moringa oleifera,* can be used in various food formulations to enhance the nutritional value and wider acceptance and consumption.

3.1 Introduction

India is one of the major contributions of agricultural produce to the world. Though the country has attained food security at the national level, hunger and malnutrition continue to haunt, despite green, white and IT revolution and economic reforms. Most imperatively, the country has to attain food security at the household level. The key reasons stated for such improper food security among Indian households are post-harvest losses, lack of awareness on healthy foods and poor utilization of natural resources. Among the plant foods, green leafy vegetables are rich in micronutrients and could be used to

S. Amutha (✉) · K. Jothilakshmi · J. Devi Priya · G. Hemalatha
Community Science College and Research Institute, Tamil Nadu Agricultural University, Madurai, India
e-mail: deanhscmdu@tnau.ac.in

N. M. Boopathi et al. (eds.), *The Moringa Genome*, Compendium of Plant Genomes,
https://doi.org/10.1007/978-3-030-80956-0_3

prepare healthy and nutritious foods to enhance the nutrient content and to attain food and nutritional security in an affordable way (Ulian et al. 2020).

India is the prime producer of Moringa (Drumstick) with an annual production of 2.2 million tonnes of tender fruits from an area of 43,600 ha. Moringa leaves have been reported to be a rich source of β-carotene, protein, vitamin C, calcium and potassium. Besides, it acts as a good source of natural antioxidants such as ascorbic acid, flavanoids, phenoloics and carotenoids and thus enhances the shelf-life of fat-containing foods.

With an annual growth rate of 26–30%, Moringa leaf export is a tender business in India especially in Tamil Nadu, Andhra Pradesh, Karnataka and Odisha. Interestingly, 80% of the production of Moringa leaves in India, fetching higher share of foreign exchange to the country. The major countries which import Moringa leaves from India are China, the USA, Germany, Canada, South Korea and several European countries. India exported Moringa leaves worth Rs. 14.6 crores in 2015 compared to Rs. 11.61 crores in 2014 and the export market has been increasing at the rate of over 30% (Venkatesan et al. 2018). Currently, Moringa leaves are priced at Rs. 70–80 per kg and that of Moringa seed costs at Rs 500 per kg.

Moringa leaves have several folds of applications, especially in the nutraceutical, pharmaceutical and cosmetics industries. Apart from leaves, Moringa seeds and oil also have a huge overseas market. The oil from the plant can be used as a biofuel.

Consumers are looking for variety of foods in their diet. It is the right time to develop nutritious and organoleptically acceptable products with locally available food sources. Development of ready to use and ready to eat foods using Moringa leaves would be the most suitable way to bring dietary improvement to the people and educating the public about the beneficial value of Moringa leaves needs due to importance. The Moringa leaf-based foods should be made available in the form of convenient foodstuff to meet the requirements of the different consumers.

The food industries, entrepreneurs and self-help groups need to promote value-added products through small and medium scale enterprises. It creates new opportunities for employment and it encourages the small farmers to go for the cultivation of Moringa leaves on a large scale. Promotion of Moringa leaf incorporated foods would help not only for alleviating micronutrient deficiencies but also for the development of functional foods using Moringa leaves helps to cure several chronic degenerative disorders. This could provide an additional source of income, employment and exports to the farmers and entrepreneurs and processors.

Hence, dehydration of Moringa leaves by reducing the moisture content below a critical level at which enzyme activity and growth of microorganisms are hindered would help to preserve Moringa leaves fairly for a longer period of time. Prior to the dehydration process, the Moringa leaves are needed to be subjected to several treatments for a successful dehydration process. Pre-treatments and drying methods check the undesirable physicochemical and enzymatic changes that may occur during drying and help to enhance the keeping quality of dried products (Johns 2007). Dehydrated leafy vegetables are simple to use and have longer shelf life than fresh vegetables (Putriani et al. 2020).

Several attempts have been made to explore the possibilities of using dehydrated Moringa leaf powder to enrich a variety of Ready-to-Eat foods, which can increase the nutritional profile of the food supplements used to alleviate nutritional hunger. This chapter provides a case study that analyses the properties of Moringa leaves when exposed to different processing methods and how to incorporate Moringa leaf powder in new food formulations.

3.2 Characteristics of Raw Moringa Leaves that Fit for Processing

It is always advisable to procure good quality dark green and fresh Moringa leaves. The procured leaves should be cleaned properly by removing the insect-damaged leaves, deteriorated leaves, stems. Initially, fresh, green Moringa leaves should be separated by discarding the thin branches, the discolored, decayed and wilted leaves since the decayed and wilted leaves gives a bad flavor to the whole batch and also these can lead to loss of nutrients too.

The leaves should be washed thoroughly with plenty of water to remove all the adhering dust and dirt particles and they should be allowed to drain the excess water. The residual moisture may be evaporated at room temperature (at 32 °C for 2–3 h), by spreading the leaves on the filter paper with constant turning over to avert the fungal growth. Such cleaned leaves alone should be used for drying and for the development of Moringa leaf-based value-added products.

3.3 Dehydration of Moringa Leaves

The cleaned Moringa leaves can be dried using several types of drying methods such as shade drying, cabinet drying and fluidized bed drying. However, before starting the drying process, the leaves may be blanched to increase the efficiency of drying. The cleaned leaves can be blanched by employing a steam blanching method that exposes the leaves for 3 min.

The major advantage of blanching is that it helps to kill the harmful microorganisms that are present on and inside the leaf, it softens the cell walls and speeds up the drying process. Besides, this process also deactivates the plant enzymes (particularly lipoxidases) that can spoil the flavour and the nutritional quality of the leaves. Blanching helps to store the dried leaf powder for several months. Hence, in the present study, the Moringa leaves were blanched for 3 min and dried. Blanching was done for the selected drying methods like shade drying, cabinet drying and fluidized bed drying.

It is also advisable to use a uniform quantity of leaves for drying as it will improve the uniform recovery of dried leaves. Observation of other parameters such as weight of the fresh leaves, temperature, time taken for drying and weight of the leaves after drying would also be useful to design an efficient drying procedure. Upon completion of the drying process, the dried leaves can be powdered and sieved before using them for the standardization of different value-added products.

3.4 Standardization of Value-Added Moringa Leaf Products

Our expertise in exploring different drying methods has indicated that blanched and shade dried leaves preserve better nutrients both during the initial stage and long-term storage. Therefore, it is better to use Moringa shade dried leaves for developing value-added food products. Nutritious products can be developed by incorporating shade-dried Moringa leaf powder with fine powders of millets, pulses and other food grains at different proportions to prepare ready to use/cook or ready to eat foods and Moringa bakery diets.

3.5 Organoleptic Evaluation of the Value-Added Moringa Products

Once the value-added Moringa products were developed, they can be subjected to organoleptic evaluation to assess the acceptability of the products using a nine-point hedonic scale: a score of one indicated that the recipe was "disliked extremely" and a score of nine denoted "liked extremely well". The same products may be dummy numbered and can be evaluated by the consistent panels of judges on different days to assess the colour, appearance, flavor, texture, taste and overall acceptability to ensure that there is no bias during the evaluation.

	9	7	8	6	5	4	3	2	1
Score card									
	Disliked extremely								Liked extremely well

3.6 Nutritional Analysis of Dried Moringa Leaves and Their Value-Added Products

To estimate the biochemical and nutritional composition of value-added Moringa products and dried Moringa leaves, the following analysis can be conducted:

- Estimation of moisture and protein content (AOAC 1995)
- Estimation of crude fibre, β-Carotene and minerals such as calcium, phosphorus, iron, potassium, sulphur, magnesium and zinc (Sadasivam and Manickam 1996 and references therein).

The data obtained from the various experiments should be subjected to statistical analysis to find out the impact of different treatments and packaging materials on Moringa leaf products.

3.7 Impact of Drying Methods on Moringa Leaves

Shade drying involves spreading the Moringa leaves after the washing and blanching step (see Sect. 3.3) on a cleaned dried tray and covered with cleaned muslin cloth and the trays are kept at room temperature (32 °C for 48–50 h). It is always preferable to have well-ventilated room for shade drying of leaves.

Under the cabinet drying procedure, the blanched leaves are loaded on the trays by forming a single layer and are dried in the dehydrator using the force air technique. The oven is preheated to 60 °C and then the loaded trays with leaves are kept in the oven. The temperature is maintained at 60 °C for 3–4 h. Several studies have suggested that drying of leaves at 60 °C yields better quality of dried leaves (Joshi and Mehta 2010; Lakshmi and Vimala 2000).

When the Moringa leaves are dried in fluidized bed drying methods, the leaves are dried under the clean air that passes over the electric heaters. The hot air passes over the product and fluidization dries the product. It is a quick-drying method. The leaves are dried at 110/80 °C for 35–45 min. A summary of drying Moringa leaves using different methods is given in Fig. 3.1.

Table 3.1 provides the details of variations in the recovery of dried Moringa leaf powder when three different drying methods were employed.

Thus, it can be concluded that though there is not much difference among the different drying methods and procedures employed to recover Moringa leaf powder (MLP). From this study, it was found that one kg of fresh leaves yields 230–280 g of dried leaves and 230–270 g of MLP depending upon the drying method and procedure. Hence, it can be concluded that five g of dried leaves or powder is approximately equal to 20 g of fresh Moringa leaves.

Table 3.2 summarizes nutrient content in fresh and dried Moringa leaves that were processed with different drying methods and procedures and clearly indicate that drying procedures have a definite impact on the nutrient content of the dried Moringa leaf powder.

3.8 Shade Drying Preserves Maximum Nutrients in Dried Moringa Leaf Powder

Invariably, it has been found that the nutrient content of dried Moringa leaves and powder prepared after dehydration became a concentrated

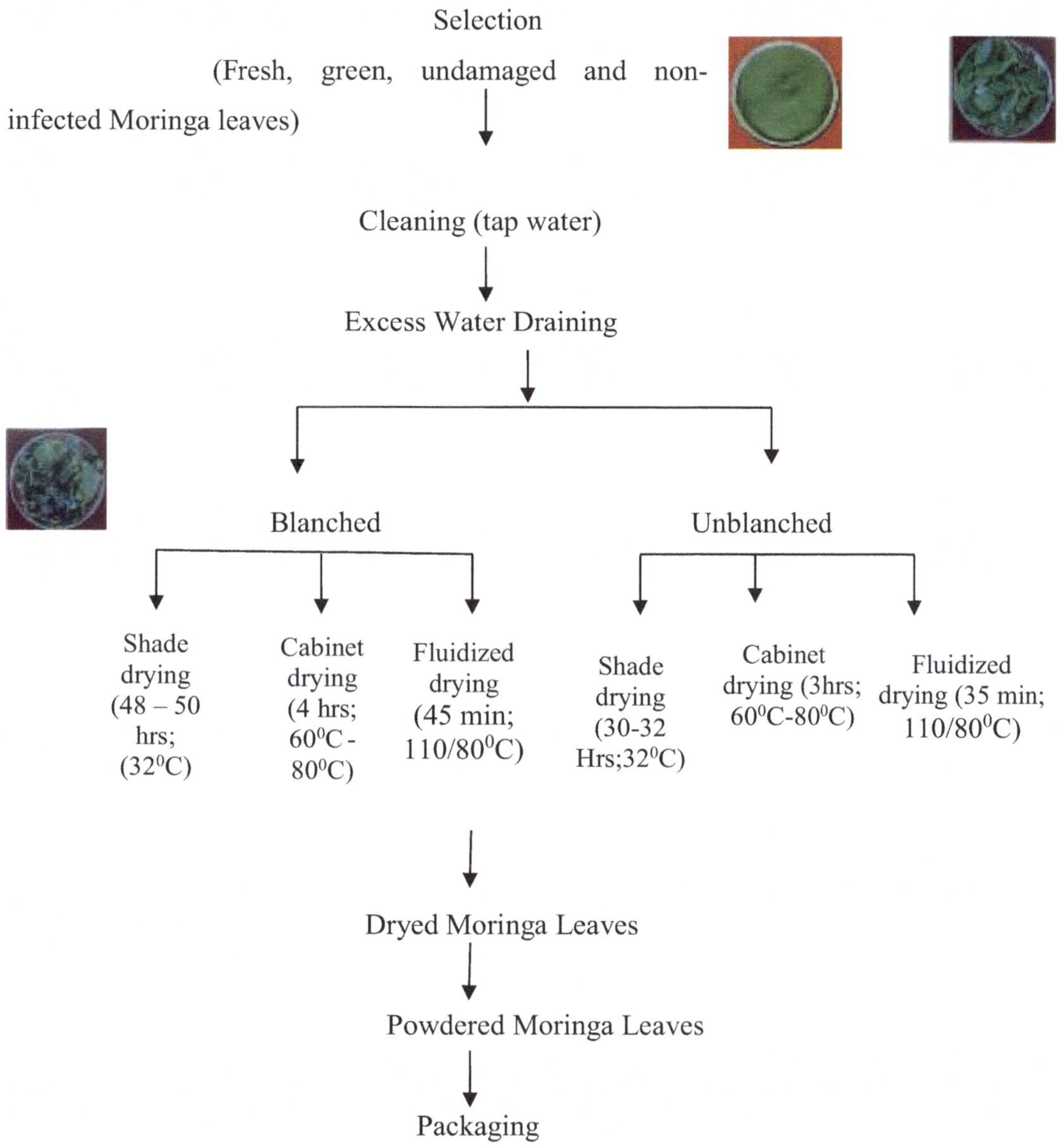

Fig. 3.1 Drying Moringa leaves using different methods

Table 3.1 Variations in the recovery of dried Moringa leaf powder

Details	Shade drying		Cabinet drying		Fluidized bed drying	
	Blanched	Unblanched	Blanched	Unblanched	Blanched	Unblanched
Initial weight (g) (after cleaning)	100.0	100.0	100.0	100.0	100.0	100.0
Dried leaves (g)	28.0	27.0	24.0	23.0	28.0	28.0
Moringa leaves powder (g)	27.0	26.0	24.0	23.0	27.0	27.0

Table 3.2 Nutrient content of fresh as well as dried Moringa leaf powder (that were processed with different methods)

Details	Fresh leaves	Shade drying		Cabinet drying		Fluidized bed drying	
		Blanched	Un blanched	Blanched	Un blanched	Blanched	Un blanched
Moisture (%)	75	5.0	5.1	5.0	5.2	5.0	5.1
Protein (g)	6.9	17	16.8	16.2	15.8	16.6	16.4
Fat (g)	1.5	3.8	3.4	3.1	3.0	3.3	3.5
Crude fibre (g)	0.8	3.8	3.6	3.1	3.0	3.5	3.4
β-carotene (μg)	19305	19,622	19,464	19,390	19,332	19,510	19,454
Vitamin-C (mg)	205	120	126	101	109	110	115
Ash (g)	2.8	3.0	3.0	3.0	3.04	3.0	2.8
Calcium (mg)	335	966	950	869	852	944	920
Phosphorus (mg)	70	210	200	194	190	198	195
Iron (mg)	1.9	2.6	3.2	2.8	2.6	3.1	3.3
Copper (ppm)	0.07	23.2	19.8	28.5	20.6	36.9	31.1
Magnesium (mg)	42	345	268	307	266	305	267
Potassium (mg)	259	101	99	100	97	107	99
Sulphur (mg)	1081.8	716	797	683	832	752	850

source of all the nutrients. The dried leaves powder retained appreciable amounts of protein, fat, crude fibre, β-carotene, calcium, phosphorus and iron content. The maximum retention of Vitamin C content observed in shade dried samples when compared to the cabinet and fluidized drying method. This is because shade dried samples are kept at room temperature, whereas heat liable vitamin C is destroyed due to oxidation when exposed to high temperature under cabinet and fluidized bed drying methods. Similar to this report, Mbah et al. (2012) evaluated the proximate and micronutrients composition of sun-dried, shade-dried and oven-dried *Moringa oleifera* leaves and also shown that that all drying techniques improved the protein, fiber, carbohydrate, vitamin A, calcium and zinc level of the leaf.

In another study, Mutiara et al. (2012) examined the effect of blanching methods and duration on the preservation of nutrients in *Moringa oleifera* leaf powder. Unblanched Moringa leaf powder contained 340 mg vitamin C, 16.51 mg beta-carotene and 24.59% crude protein in 100 g dry mass. On the other hand, the blanched *Moringa oleifera* leaves resulted in a decrease in the level of vitamin C (120–238 mg in 100 g^{-1} dry mass), increase in the level of beta carotene (19.36–21.52 mg in 100 g^{-1} dry mass) and increase in the level of protein 24.70–30.68%. Therefore, it can be concluded that blanching can be used to increase the beta carotene and protein but not vitamin C.

Shadow drying though took longer than sun and oven drying, lead to less loss of carotene. The maximum retention of beta carotene content was in shade-dried sample (39600 μg) was followed by oven dried sample (37800 μg) and a minimum level of 36000 μg/100 g leaf powder in sun-dried sample. The fibre content in the three methods was in the range of 11.3–12.1% with the highest level in shadow dried sample followed by oven dried. In contrast, minimum values of fibre were found in sun-dried Moringa leaf sample in other studies (Joshi and Mehta 2010).

Value addition technologies using moringa leaves powder

Moringa leaf Based Ready to Use Foods	**Moringa leaf Based Ready To Eat Foods**
(i) Breakfast foods Moringa leaf adai mix, Moringa leaf snack mix, Moringa nutri roti mix, Moringa spiced chappathi mix, Moringa versatile food mix, Moringalicks **(ii) Spiced mixes** Moringa nutri sambar mix, Moringa nutri rasam mix, Moringa juice mix, Moringa rice mix, Moringa ready to use soup mix, Ready to use moringa chutney mix, Moringa leaf paruppu podi mix	**(i)Traditional ready to eat foods** Moringa leaf pittu, kollukattai, boli, dosa, porridge, nutritious ball, roti, Kolaiputtu **(ii) Moringa leaf bakery products** Moringa nutri rich cookies (sweet, spicy and coconut), moringa puddings **(iii) Moringa health beverages** Moringa Malt

Fig. 3.2 Development of different valued-added products using Moringa leaf powder

3.9 Value Addition Technologies for Moringa Leaf Powder

Nutritious value-added Moringa products can be developed by incorporating millets and pulses for the development of Ready-To-Use/Cook foods and Ready-To-Eat foods. The process of developing different value-added products is outlined in Fig. 3.2.

3.10 Moringa Leaf Value-Added Products and Its Nutritional Aspects

Organoleptic evaluation of a range of Moringa leaf value added was found to be good for all the tested products (Fig. 3.3) and they are well received by the panelists. The bitter taste of drumstick leaf (which was usually found when the Moringa leaves are consumed without any processing) was lost during the processing methods and the value-added products were rated high ranks. The sensory analysis shows the acceptability of all the Moringa foods in terms of colour, flavor, texture, taste and overall acceptability in the range of 8.0–8.8 in a 9-scale scoring.

The analysis of Moringa value-added foods indicated that it contains more than sufficient amounts nutrients such as protein, β-carotene, calcium and iron and the details of the nutritional profile of each product is provided in Table 3.3.

Moringa value-added products can be prepared by adding three to ten per cent of dried Moringa leaf powder in 100 g of staple food which is equal to the consumption of 15–50 g of fresh Moringa leaves. It has also been found that such value-added food supplements are preferred by all age groups as this little quantity of Moringa easily digestible. Carbohydrates are energy providers that get converted into glucose in our bodies during digestion and it is a main source of fuel for our body and it is especially important

Fig. 3.3 Value-added products developed using Moringa leaf powder

for regular brain function. From Table 3.3, it is evident that 100 grams of Moringa leaf value-added foods provided 40–60 g of carbohydrate (depending on the type of value-added product) and supplied enough energy to our daily activities. The highest protein content 18.64 g was found in Moring leaf-paruppu podi, which was prepared by mixing different pulse flours and Moringa leaf powder. This dietary protein is essential to supply amino acids for the growth and maintenance of cells and tissues.

The β-carotene content was ranged from 1062 μg to 17,268 μg in 100 g of Moringa foods; thus, it can be the best food supplement to school children who have Vitamin A deficiency. It was estimated by the National Nutrition Monitoring Bureau (NNMB) that the prevalence of Vitamin A deficiency is 61% at the national level, 49% in Tamil Nadu and 74% among the Indian tribal population (Laxmaiah 2006). Hence, the intake of β-carotene-rich Moringa value-added foods in their regular diet would help to prevent Vitamin A deficiency.

The highest mineral contents of calcium, phosphorus and iron of Moringa leaf value-added foods were ranged from 124.6 mg to 241.5 mg, 63 mg to 288.64 mg and 0.8 mg to 4.71 mg, respectively (Table 3.3) and it was chiefly due to the incorporation of Moringa leaf powder in the food products.

In general, cookies are prepared with refined wheat flour and they are a popular snack item. With the health-conscious initiative, the cookies have been prepared in this study with whole wheat flour, millet flour and Moringa leaf powder. Such cookies contained 11.2% of protein, 5.1% of fat, 1320 μg of β-carotene, 572 mg of calcium, 256 mg of phosphorus and 3.0 mg of iron/100 g of cookies. In general, cookies are preferred by all age groups and hence the administration of such exceedingly nutritive cookies would not be a problem. These Moringa value-added cookies fulfilled the daily nutritional requirement of 1-to 3-year-old children such as protein (22 g/day), iron (12 mg/day), calcium (400 mg/day) and β-carotene (1600 μg/day) (Gopalan et al. 2002).

Jagadeesan et al. (2020) studied the proximate composition, minerals (Ca, Fe, K, P) and antioxidant scavenging activity of Moringa Leaf Powder. In total, six different instant value-added products were formulated by adding the developed MLP, which were then evaluated in comparison to the control samples. The protein content was about 23% in MLP. The MLP was a

Table 3.3 Nutritional composition of value-added products prepared from Moringa leaves

Nutrients	Moringa leaf adai mix,	Moringa leaf snack mix,	Moringa leaf nutri roti mix	Moringa spiced chappathi mix	Versatile food mix.	Moringalicks	Moringa nutri sambar mix	Moringa nutri rasam mix	Moringa ready to use soup mix	Moringa juice mix	Moringa rice mix	Ready to use Moringa chutney mix	Moringa leaf paruppu podi mix
Moisture (%)	5.3	5.2	5.1	6.3	5.5	6.1	7.3	7.4	6.0	6.1	5.8	7.5	5.98
Carbohydrate (g)	59	37	60.7	66.6	58.3	64.8	36.5	42.6	40.3	20.8	41.9	*40.6*	*48.9*
Protein (g)	5.7	6.3	5.6	6.6	6.5	6.8	2.3	2.9	18.60	6.9	21.7	4.5	18.64
Crude fibre (g)	4.0 g	2.7 g	3.6	3.2	2.3	5.5	3.6	3.8 g	4.6	4.0	2.9	3.4	1.44
β-carotene (μg)	1758	1913	1062	7364	6358	1893	1736	1949	3337	17,268	5108	*2144*	1277.04
Calcium (mg)	154.2	124.6	197.6	167.6	141.5	146.75	174	241.5	213	217.7	197	213	221.55
Phosphorus (mg)	75	63	187.2	237.2	138.8	312.8	196	52.5	202	190.50	201	202	288.64
Iron (mg)	3.3	2.3	3.6	4.6	4.2	4.3	3.0	0.8	2.4	3.1	2.4	2.4	4.71

rich source of calcium and iron (8616 mg and 659 mg, respectively) besides, other micronutrients and they were shown to be highly preferred for consumption (Ghosh et al. 2020).

3.11 Conclusion

Moringa oleifera is an excellent source of macro and micronutrients including antioxidants. However, Moringa leaves are not as popular as that of other leafves such as spinach, and fenugreek; but they are used in local preparation only in Southern and Eastern parts of India though Moringa has enormous potential. It is generally agreed that there is a knowledge gap in the potential uses of Moringa and its use in food fortification. It can be utilized to make foods that could be a step towards curbing malnutrition through the development of innovative valued-added products that would be easily acceptable to the consumers. The high levels of vegetarian diet in the Indian population (which may not provide necessary protein and micronutrients, if not selected appropriately) and the elevated malnutrition level at the rural and urban population drives such research, wherein an inexpensive, abundantly available, highly nutritious and ethnic food like the *Moringa* leaves can address gaps in the nutritional needs. These products can be consumed by all age groups and can be eaten as an addition to the main course. Hence, such value-added and innovative nutritious dishes can support people to eat healthily and also continue to eat traditional and local food that has not consumed a large carbon path.

References

AOAC (1995) Official methods of analysis. 14th Editon. The Association of official Agricultural Chemists. Washington, D. C

Ghosh P, Chatterjee S (2020) Evaluation of organoleptic, proximate parameters and analysis of nutritional composition of five wild weeds: a search for low-cost nutraceuticals. Int J Pharmaceutical Sci Res 11 (10):5170–5181

Gopalan C, Ramasastri BV, Balasubramanian SC (2002) Nutritive value of indian foods. National Institute Nutrition, Hyderabad, India, pp 1–156

Johns T (2007) Agrobiodiversity, diet and human health. In: Jarvis DI, Padoch C, Cooper HD (eds) Managing biodiversity in agricultural ecosystems. New York: Columbia University Press, p 492

Joshi P, Mehta D (2010) Effect of dehydration on the nutritive value of drumstick leaves. J Metabolomics Syst Biol 1(1):6–9

Lakshmi B, Vimala V (2000) Nutritive value of dehydrated green leafy powders. J Food Sci Technol 37 (5):465–471

Laxmaiah A (2006) Prevalence of vitamin a deficiency among preschool children in rural India. Report No.: NNMB Technical Report No 23. Hyderabad: National Institute of Nutrition; Available from: http://www.ninindia.org/NNMB-PDF%20FILES/VAD-REPORT-final-21Feb07.pdf. Accessed Jan 12 2018

Mbah BO, Eme PE, Paul AE (2012) Effect of drying techniques on the proximate and other nutrient composition of Moringa oleifera leaves from two areas in Eastern Nigeria. Pakistan J Nutr 11(11):1044–1048

Mutiara O, Negi PS, Roy SK (2012) Effect of drying conditions on quality of green leaves during long term storage. Food Res Int 34:283–287

Putriani N, Perdana J, Meiliana, Nugrahedi PY (2020) Effect of thermal processing on key phytochemical compounds in green leafy vegetables: a review. Food Rev Int, 1–29

Sadasivam S, Manickam A (1996) Biochemical methods. IInd edition, New Age International (P) Ltd. New Delhi: 63

Jagadeesan S, Sarangharaajan A, Ravikumar N, Palani K, Ramanathan RM (2020) Development of Ready to Use value added products from Moringa oleifera leaves. Int J Food Sci Nutr 5(2):60–62

Ulian T, Diazgranados M, Pironon S, Padulosi S, Liu U, Davies L,... Sharrock S (2020) Unlocking plant resources to support food security and promote sustainable agriculture. Plants People Planet 2 (5):421–445

Venkatesan N, Sekhar C, Murugananthi M, Vidhyavathi A (2018) Marketing and price spread analysis of moringa in Tamil Nadu, India, Horticult Int J 2 (4):212–22

4 Medicinal and Therapeutic Properties of Moringa

N. Manavalan, N. Manikanda Boopathi, and M. Raveendran

Abstract

There are many medicinal plants that have been used for thousands of years and due to the recent COVID-19 pandemic, these plants can now be found in several healthcare products because of their therapeutic efficacy. Due to its rich medicinally imperative phytochemicals, *Moringa oleifera* has attracted several health industries. Besides, Moringa has been preferred as an affordable resource to fight malnutrition owing to its abundant nutritional properties. A large array of reports has repeatedly shown that Moringa can be used as anticancer, antidiabetic, anti-inflammatory and antioxidant products. Though such references provided evidence for its potentials as a functional food, yet the basic questions on its mechanisms of action, especially on its medicinal properties remain to be unclear. This chapter summarizes the individual bioactive phytocompounds present in Moringa and their medicinal properties. It also highlights those further studies that need to investigate the likely mechanisms of actions of the Moringa phytochemicals that foster its informative applications in the prevention and management of chronic diseases.

4.1 Moringa: A Treasure Trove for Therapeutic Properties

Ancient Indians, Romans, Greeks and Egyptians have long been using Moringa (*Moringa oleifera*) as one of the medicinally imperative plant products. Evidence for this usage is found in the large arrays of traditional and tribal medicinal knowledge and non-peer-reviewed sources (Fahey 2005).

Likewise, during the past three and half decades, enormous numbers of reports have surfaced in contemporary scientific periodicals that experimentally proven its pharmaceutical, therapeutic and prophylactic properties and its economic and societal importance. Above and beyond, Moringa has also been utilized in water clarification, lubricating and cosmetics industry, timber, wood coal and natural fencing.

Recent rigorous scientific proofs have critically examined the pharmaceutical properties of Moringa leaves, flowers, pods and other parts of the tree and justified that Moringa can be utilized as a "healthy food' or "superfood". Though it cannot be a panacea as promoted by the suppliers of Moringa products, constantly accumulating scientific reports supported the therapeutic and

N. Manavalan
Government Yoga and Naturopathy Medical College, Chennai, India

N. M. Boopathi (✉) · M. Raveendran
Department of Plant Biotechnology, Centre for Plant Molecular Biology and Biotechnology, Tamil Nadu Agricultural University, Coimbatore, India
e-mail: nmboopathi@tnau.ac.in

N. M. Boopathi et al. (eds.), *The Moringa Genome*, Compendium of Plant Genomes,
https://doi.org/10.1007/978-3-030-80956-0_4

prophylactic value of the Moringa products in almost all cases.

Moringa has been advocated for a long time as "natural nutrition for the tropics" and it has been used to combat malnutrition (which is a cause of several life-threatening diseases or disorders) in infants and nursing mothers. Moringa leaves can be consumed as fresh or cooked. They can also be stored as dehydrated powder in an air-tight container for several months. Such storage does not require any refrigeration and it has been reported that there is no loss of its nutritional value during these storing procedures. Furthermore, when other food sources are insufficient during the dry season, Moringa is in its full leaf production potential at the end of the dry season and hence, it has been known as a hopeful food source in the tropics.

Moringa possesses a rich amount of several vitamins and minerals as well as carotenoids (including β-carotene or pro-vitamin A). In addition to that Moringa also has a range of unprecedented unique compounds. For example, Moringaceae is rich in glucosinolates and isothiocyanates which are phytocompounds containing the simple sugar, rhamnose (Bennett et al. 2003). The exclusive compounds that are present in Moringa include 4-(a-L-rhamnopyranosyloxy) benzyl isothiocyanate, 4-(4'-O-acetyl-a-L-rhamnopyranosyloxy) benzyl isothiocyanate, niazimicin, pterygospermin, benzyl isothiocyanate and 4-(a-L-rhamnopyranosyloxy) benzyl glucosinolates and they have been reported to have hypotensive, anticancer, and antibacterial activity (Fahey et al. 2001, 2002; Fahey 2005). The above phytocompounds are only to quote few examples; other unique and valuable compounds are provided in the subsequent sections.

4.2 Phytochemical Compounds of Moringa

Secondary metabolites that are accumulated in high concentrations in response to both biotic and abiotic environmental stimuli are referred to as phytochemicals and they are also shown to possess several medicinal properties. Since time immemorial, such phytochemicals are used as a therapeutic agent to cure as well as to prevent against various contagious and non-contagious diseases. Owing to the abundant content of biologically active phytochemical and nutritional compounds, Moringa has long been explored as health-promoting plant products and large numbers of studies have and being undertaken to understand its medicinal properties.

Plant-derived phytochemicals are grouped into five classes according to their chemical structures: polyphenols, carotenoids, alkaloids, terpenoids, and sulphur-containing compounds. Interestingly, among the plant species, Moringa has almost all of these phytochemicals and it was strongly thought that the diverse biological activities and disease preventive potential of Moringa are largely because of such phytochemicals (Ma et al. 2020).

For example, Moringa has been shown to contain rich amounts of polyphenols including flavonoids and phenolic acids. Generally, polyphenols are having either one (e.g. phenolic acids) or more than one phenol ring (e.g. flavonoids) in their chemical structure and one of the key components of phytochemicals that have medicinal properties. Among the flavonoids, myrecytin, quercetin and kaempferol are found in Moringa leaves in the concentrations of 5.8, 0.207 and 7.57 mg/g, respectively (Mohammed and Manan 2015).

It should also be noted that the concentrations of each flavonoid varied depending on which part of the Moringa plant was used for the analysis using advanced chromatography techniques such as HPLC, GC–MS and LC–MS. For example, quercetin and kaempferol glycosides (glucosides, rutinosides and malonyl glucosides) were not present in roots and seeds. However, 0.46–16.64 and 0.16–3.92 mg/g dry weight of leaf were quercetin and kaempferol, respectively (Saini et al. 2016). Similarly, a lower amount of myricetin, rutin and epicatechin were also detected in Moringa leaves.

Phenolic acids such as gallic acid, caffeic acid, chlorogenic acid, coumaric acid and ellagic acid were also found in Moringa leaf and pod. These are a sub-group of phenolic compounds that are mostly originated from hydroxybenzoic acid and hydroxycinnamic acid and shown to possess antioxidant, anti-inflammatory, antimutagenic and anticancer properties.

The antioxidant activity of phenolic compounds (such as neutralizing free radicals, quenching singlet or triplet oxygen, or decomposing peroxides) may be due to the inactivation of lipid-free radicals or owing to their redox properties that help in avoid decomposition of hydroperoxides into free radicals.

The chief phenolic acid present in Moringa is chlorogenic acid (CGA), which is an ester of dihydrocinnamic acid, which impedes glucose-6-phosphate translocase in the liver and thus reduce hepatic gluconeogenesis and glycogenolysis. Therefore, CGA has been shown to have an important role in glucose metabolism regulation.

Total phenolic concentrations in Moringa leaves were in the range of 2000–12,200 mg GAE/100 g as determined by Folin–Ciocalteau assay (Leone et al. 2015). The reason for the great variation in phenolic content is attributed to the fact that it depends on genotype, geographical location and environmental conditions. Though flowers and seeds were found to have polyphenols, they were present in relatively lesser concentrations than the leaves (Alhakmani et al. 2013).

Significant amounts of tannins (which are complex polyphenol molecules and have the ability to bind and precipitate protein, amino acids, alkaloids and other organic molecules in aqueous solutions) were also found in Moringa leaves. The highest concentration of tannins was reported in dried leaves (20.7 mg/g) and small amounts of tannins were also found in seeds (Mohammed and Manan 2015).

Different parts of Moringa plant were found to possess different types of glucosinolates (which are heterogeneous groups of sulphur and nitrogen-containing glycosidic compounds). Moringa contains 4-(α-L-rhamnopyranosiloxy) benzyl glucosinolates and is also called as glucomoringin and it has been reported in Moringa stem, flowers, pods, leaves and seeds. The maximum amount of glucomoringin was found in seeds (8620 mg/100 g) followed by leaves (78 mg/100 g) (Maldini et al. 2014).

However, the major glucosinolate found in the roots of the Moringa plant is benzyl glucosinolate (also called as glucotropaeolin) though there were reports on significant variations in its concentration which are due to geographical regions in which the Moringa plants are grown (Bennett et al. 2003). Myrosinase catabolizes glucosinolates into glucose, isothiocyanates, nitriles and thiocarbamates and these compounds were also reported in Moringa (Waterman et al. 2015).

Colours of leaves, fruits and vegetables are mainly due to carotenoids. Both green and dried leaves of Moringa contain different types of carotenoids. For example, green Moringa leaves contain 6.6–17.4 mg/100 g β-carotene (also known as pro-vitamin A) and dried leaves has 23.31–39.6 mg/100 g β-carotene, which are much higher than those found in carrots, pumpkins and apricots (Glover-Amengor et al. 2017).

It was also reported that the Moringa cultivars grown in India contain several other carotenoids too in the foliage, flowers and immature pods (fruits). More than 50% of total carotenoids was All-E-lutein and it was found in foliage and fruits. Besides, other classes of carotenoids such as All-E-luteoxanthin, 13-Z-lutein, all-E-zeaxanthin and 15-Z-b-carotene has also been reported in smaller quantities in Moringa (Saini et al. 2016).

Moreover, Moringa has also been confirmed to possess alkaloids (which are derived from amino acids metabolism) and N, α-L-rhamnopyranosyl vincosamide is the most abundantly reported indol-alkaloid in its leaves. Besides, pyrrolemarumine 4″-O-α-L-rhamnopyranoside (marumosides A) and 4′-hydroxyphenylethanamide (marumosides B) (which are unusual glycosides of a pyrol alkaloid) were also reported but their concentration in Moringa leaves are yet to be quantified (Sahakitpichan et al. 2011).

A plentiful amount of saponins in Moringa freeze-dried leaves (64–81 g/kg of dry weight) was also found. Saponins greatly minimize the absorption of cholesterol by binding bile acids and boosting fecal excretion and ultimately the lowering plasma cholesterol. Besides, saponins also prevent cancer developments.

4.3 Moringa for Disease Therapy and Deterrence

Though the voluminous folklore history highlighted the benefits of Moringa based products such as extracts, decoctions, poultices, creams, oils, emollients, salves, powders and porridges as a healthy diet or to cure or prevent disease or infection, it has long been insisted to develop a scientific basis for these claims (Talalay and Talalay 2001).

Accumulating scientific evidence have authenticated the majority of the above healthcare claims and large numbers of scientific literature support the role of Moringa products in antibiotic, antitrypanosomal, antiulcer, antibiosis, antispasmodic, anti-inflammatory, hypotensive, hypocholesterolemic, hypoglycemic activities and even reduction of *Schistosome cercariae* titer (Farooq et al. 2012 and references therein).

Fahey (2005) has provided an extensive list of references that were studied evidently the nutritional, therapeutic and prophylactic applications of Moringa. Recent literature survey on the above line has also been witnessed all-inclusive list of studies that support the utility of Moringa as a medicinal plant (Farooq et al. 2012; Mehwish et al. 2020; Meireles et al. 2020; Padayachee and Baijnath 2020). The below subsections provide few representational references that support several therapeutic and prophylactic properties of Moringa.

But it should also be kept in mind that amidst of in vitro (cultured cells) and in vivo (animal) trials that offered systematic support benefits of Moringa-based traditional medicine lore, it has been underlined on several occasions that none of the experiments has clearly shown their efficacy in human beings. Especially there is a dearth of controlled and randomized clinical experiments and experimental findings published in peer-reviewed journals.

4.3.1 Antibiotic (or Anti-hyperglycemic) Activity

Indeed, the majority of the reports have pinpointed the antibiotic ability of Moringa and few studies have shown the Moringa phytochemicals that possess antibiotic behaviour (Gopalakrishnan et al. 2016). For example, antibiotic activity of 4-(α-L-rhamnopyranosyloxy) benzyl glucosinolate and its cognate isothiocyanate against several species of bacteria and fungi have been reported (Eilert et al. 1981).

4.3.2 Anticancer Activity

Numerous recent reviews have provided a list of studies that have reported the anticancer activities of Moringa based products and they were proved to prevent the development of human cancer cells (Ma et al. 2020; Szlachetka et al. 2020; Sodvadiya et al. 2020; Gupta et al. 2020 and references therein). It has been evidently shown that extracts of Moringa leaves, bark and seed have arrested the development of breast (MDA-MB-231) and colorectal (HCT-8) cancer cell lines of humans by reducing cell motility and colony formation and resulted in low cell survival, high apoptosis and enrichment of G2/M.

Similarly, increased apoptosis, DNA fragmentation and oxidative stress were reported when aqueous extract of Moringa leaves was used to treat human cancerous lung cells (A549) and other types of cancer cells and thereby prevent cancer proliferation and invasion.

Dose-dependent inhibition (having both antiproliferative and apoptosis properties) of cell proliferation of human tumor KB cells was also noticed for the Moringa leaf extract by stimulating apoptosis, morphological changes and DNA fragmentation. Likewise, the growth of human cancer pancreatic cells such as Panc-1, p34, and COLO 357 was curtailed by 0.75 mg/ml aqueous extract of Moringa leaves by inhibiting nuclear factor kappa B signaling pathway proteins and improved the cisplatin chemotherapy efficiency. In addition, cytotoxic effect on human peripheral blood mononuclear cells and nil effect on the hemolytic activity of erythrocytes has also been reported when aqueous extract of Moringa seeds was administrated.

Thus, it can be concluded that extracts from Moringa can be used as a therapeutic as well as preventive agent to significantly limit the progression and invasion of human cancer cells (such as myeloma, cervix, colon, breast, leukemia, lung, liver, neuroblastoma, pancreas, colorectal, epidermoid, oral, ovarian, muscular, prostate, skin).

4.3.3 Antioxidant Activity

Antioxidant properties of extracts obtained from Moringa leaves, seeds and pods have also been established in several studies (reviewed in Ma et al. 2020). For example, hydroethanolic extract of Moringa leaves was investigated in an in vitro study and found that there was a strong antioxidant activity: use of nitric oxide (NO) scavenging assay recorded IC50 of 120 μg/ml and deoxyribose degradation assay recorded IC50 of 178 μg/ml. Further, it was established that the presence of total phenolic and flavonoid content, carotenoids, lycopene, ascorbic acid and anthocyanins in the Moringa leaves might be responsible for the antioxidant property (Vats and Gupta 2017).

Applicability of antioxidant activity of Moringa in the biological system was further confirmed by nourishing carbon tetrachloride (CCl_4)-intoxicated rats with Moringa leaves (50–100 mg/day) for 15 days significantly reduced the lipid peroxides and improved glutathione levels besides reducing the enzyme kinetics of superoxide dismutase and catalase in the liver and kidney, when compared with control. Similar kind of trends was also observed when correlating the antioxidant efficiency of Moringa leaves (100 mg/dl per day) and showed a comparable effect to a group treated with a standard treatment of vitamin E at 50 mg/dL per day (Verma et al. 2009). In the same lines, feeding goats regularly with Moringa leaf powder increased antioxidant enzymes and decreased peroxidation in the liver of goats (Moyo et al. 2012). Correspondingly, daily uptake of 7 g of Moringa leaf powder by postmenopausal women continuously for 3 months have shown to increase antioxidant enzymes, ascorbic acid, serum levels of retinol but reduced serum malondiadehyde (Kushwaha et al. 2014a).

The aforesaid studies clearly indicate that Moringa has robust antioxidant activities and can be effectively used to cure oxidative stress-induced diseases or maladies.

4.3.4 Anti-inflammatory Activity

Pro-inflammatory mediators secreted by lipopolysaccharide (LPS)-induced murine macrophage cells were used to examine the anti-inflammatory property of ethanolic extract of Moringa pods. Increased concentrations of Moringa extract have shown an increased degree of inhibition on mRNA expression as well as the levels of interleukine-6 (IL-6), tumor necrosis factor-alpha (TNF-α), inducible nitric oxide synthase (iNOS) and cyclooxygenease-2 which may be partly mediated by preventing phosphorylation of inhibitor kappa B protein and mitogen-activated protein kinases (Muangnoi et al. 2012).

Similarly, using the RAW macrophage cell system, the negative impact of Moringa leaf concentrate and isothiocyanates on the gene expression and production of inflammatory markers (viz., iNOS, IL-1β, nitric oxide (NO) and TNF-α) has been demonstrated (Waterman et al. 2015).

In a recent study, Yan et al. (2020) have employed a lipopolysaccharide (LPS)-induced RAW 264.7 cell model to evaluate the Moringa leaf acetone extract for its anti-inflammatory effect and found that it significantly inhibited NO production and inducible NO synthase (iNOS) mRNA levels. Increased dose of Moringa extract has increased inhibitory activity and such activity was attributed to its phenolic compounds (including quercetin derivatives, kaempferol, chlorogenic acid, isothiocyanates, and kaempferol derivatives).

4.3.5 Hepatoprotective Activity

By reducing tissue histopathology, aspartate aminotransferase, alkaline phosphatase, alanine aminotransferase and lipid peroxidation and increasing glutathione, Moringa leaf extracts have been proven to protect the liver from the oxidative damage (Das et al. 2012). Liver damage induced by antitubercular drugs (such as isoniazid, rifampicin, and pyrazinamide) in rats has also been safeguarded by ethanolic extract of Moringa leaves through hepatoprotective activity by diminishing the serum levels of aspartate aminotransferase, alkaline phosphatase, alanine aminotransferase and bilirubin and by inhibiting lipid peroxidation in the liver (Pari and Kumar 2002).

In another study, Artesunate-amodiaquine (an antimalarial drug) induced liver injury in Wistar rats was studied and hepatoprotective activity of Moringa leaf extract was reported as it reduced serum AST values and hepatocyte degeneration in Artesunate-amodiaquine intoxicated rats (Okumu et al. 2017).

4.3.6 Antidiabetic Activity

Elevated concentration of blood glucose leads to a metabolic disorder, diabetes and uncontrollable glucose concentration beyond the recommended ranges will lead to health complications. As recently indicated by International Diabetes Federation (IDF) globally more than 366 million people are under diabetes mellitus and by 2030, this figure may shoots-up to 552 million or even more.

There are two kinds of diabetes mellitus and type II is characterized by abnormal glucose tolerance due to insufficient insulin secretion. Hyperglycemia is metabolic disturbance due to deficient insulin and it is a universal health concern as it causes microvascular complications (retinopathy, nephropathy and neuropathy) and macro-vascular complications, i.e. cardiovascular comorbidities leading to insulin resistance disorder.

Ma et al. (2020 and references therein) have listed out experiments that support the antidiabetic properties of Moringa. *M. oleifera* leaves were reported to have an extreme inhibitory action on α- amylase and α-glucosidase activities (Jaiswal et al. 2009) and anti-hyperglycemic activity (William et al. 1993). The key phytocompounds that attributed to this significant reduction in blood glucose level are glycosides, anthocyanins, anthraquinone, hemlock tannin and A-phenolic steroids. *M. oleifera* leaves were repeatedly shown to contain these compounds in significant amounts.

Recently, an *in silico* analysis of Anthraquinone, Sitogluside (glycoside), Hemlock Tannin, A-Phenolic Steroid, 2-Phenylchromenylium (Anthocyanins) has indicated that they were extremely selective and efficiently intermingled with mutated protein of diabetes (Zainab et al. 2020).

In supporting the antidiabetic property of Moringa, a study conducted by Kushwaha et al. (2014b) has concluded that routine consumption of Moringa leaf powder (7 g per day) by the post-menopausal women for 3 months has considerably lowered the fasting blood glucose to 13.5% and it has also been shown that such glucose-lowering effect of Moringa is due to blocking intestinal glucose, better insulin secretion and reduction in insulin resistance (Muhammad et al. 2016).

Though such studies provided evidence for the Moringa's potential in lowering the blood glucose potential, almost all of these studies were conducted in animals, and only a few studies have focused on humans. Therefore, thorough and long-term randomized controlled investigations on the antidiabetic effect of Moringa leaves, seed and pods in humans are warranted.

4.3.7 Anticardiac Arrest Activity

Investigations on the hypolipidemic effect of feeding Moringa leaves (100 mg/kg of body weight (bw)) to rabbits significantly lowered cholesterol levels by 50% and 86.52% reduction in atherosclerotic plaque formation in internal carotid and it has been observed that such hypolipidemic effects were on par with simvastatin (reference drug) treated group (Chumark et al. 2008).

In another study, lowering of serum cholesterol by 14.35% was noticed due to continuous consumption of aqueous extract of Moringa leaves (1 g/kg bw) for 30 days by rats and also lowered the concentrations of cholesterol by 6.40% and 11.09% in the liver and kidney, respectively (Ghasi et al. 2000).

Similarly, decreased levels of serum lipids in a dose-dependent manner were noticed in the rats that were fed with methanolic extracts of Moringa leaves at different doses (150, 300, or 600 mg/kg bw) for 30 days and the Moringa diet was shown to significantly reduce the triglyceride, cholesterol, low-density lipoprotein (LDL), very-low-density lipoprotein (VLDL), atherogenic index and increase the high-density lipoprotein (HDL) (Jain et al. 2010).

Few reports have also highlighted the antihypertensive property of Moringa in the rat (daily feeding @ 30 mg/kg of bw) and it was shown to be the presence of thiorcarbamate, isothiocyanate glucosides and hydroxybenzoate in the Moringa leaf extracts (Faizi et al. 1998).

Despite the above proofs, it is generally believed that more and critically acclaimed clinical experiments, with animal as well as human, are required to validate the role of Moringa in curing cardiovascular diseases and its antihypertensive activity.

4.3.8 Anti-ocular Disorder Activity

As both leaves and pods of Moringa are shown to be rich in Vitamin A, it can also be used to avert night blindness and eye problems. Deficiency in Vitamin A generally leads to blindness, which ranges from impaired dark adaptation to night blindness. It has been evidently shown that consuming Moringa leaves supplemented with oils (including ben oil, extracted from Moringa seeds) significantly improved the Vitamin A nutrition to the human and hindered the onset of night blindness and/or cataracts (Yan et al. 2020).

4.3.9 Immunomodulatory Activity

Sudha et al. (2010) investigated cellular immunity and humoral immunity in animals by exploring immunomodulatory action of methanolic extract of *Moringa oleifera* (MEMO). It was found that there was a significant increase in serum immunoglobulins and a decrease in mortality due to bovine *Pasteurella multocida* in mice. It has also been shown that MEMO induced substantial rise in adhesion of neutrophils, reduction of cyclophosphamide induced neutropenia but an increase in phagocytic index. Thus, it has been established that Moringa stimulates both cellular and humoral immunity with a low dose of MEMO (Sudha et al. 2010). Several other studies have also been recognized the immunoregulatory activity of Moringa (Mehwish et al. 2020 and references therein).

4.4 Concluding Remarks

Though the mechanism of actions was unknown, Moringa has shown to exhibit several medicinal, therapeutic and prophylactic properties due to their rich nutritional and curative compounds. Further clinical and advanced biochemical studies are required to unravel the mechanism of actions of these phytochemicals and more convincing investigations are needed to declare Moringa as a functional food to prevent and/or manage chronic diseases.

Considering the medicinal and therapeutic potentials of Moringa, Government of Tamil Nadu, India, is supplying Moringa leaf powder to the patients through Yoga and Naturopathy doctors serving in Government Medical College Hospitals and Hospitals located at Headquarters of each District and Taluk, Primary Health Centers, Yoga and Naturopathy Maternity Clinics, and Wellness Centers. An initial survey has shown that the Moringa products are well received by the patients and positive feedback on the health benefits of Moringa has been obtained so far.

References

Alhakmani F, Kumar S, Khan SA (2013) Estimation of total phenolic content, in-vitro antioxidant and anti-inflammatory activity of flowers of Moringa oleifera. Asian Pacific J Trop Biomed 3:623–627

Bennett RN, Mellon FA, Foidl N, Pratt JH, Dupont MS, Perkins L, Kroon PA (2003) Profiling glucosinolates and phenolics in vegetative and reproductive tissues of the multi-purpose trees Moringa oleifera L. (horse-radish tree) and Moringa stenopetala L. J Agric Food Chem 51:3546–3553

Chumark P, Khunawat P, Sanvarinda Y, Phornchirasilp S, Morales NP, Phivthong-Ngam L, Ratanachamnong P, Srisawat S, Pongrapeeporn KU (2008) The in vitro and ex vivo antioxidant properties, hypolipidaemic and antiatherosclerotic activities of water extract of Moringa oleifera Lam Leaves. J Ethnopharmacol 116:439–446

Das N, Sikder K, Ghosh S, Fromenty B, Dey S (2012) Moringa oleifera Lam leaf extract prevents early liver injury and restores antioxidant status in mice fed with high-fat diet. Indian J Exp Biol 50:404–412

Eilert U, Wolters B, Nahrstedt A (1981) The antibiotic principle of seeds of Moringa oleifera and Moringa stenopetala. Planta Med 42:55–61

Fahey JW, Zalcmann AT, Talalay P (2001) The chemical diversity and distribution of glucosinolates and isothiocyanates among plants. Phytochemistry 56(1):5–51

Fahey JW, X Haristoy, PM Dolan, TW Kensler, I Scholtus, KK Stephenson, P Talalay, and A Lozniewski (2002) Sulforaphane inhibits extracellular, intracellular, and antibiotic-resistant strains of Helicobacter pylori and prevents benzo[a]pyrene-induced stomach tumors. Proceedings of the National Academy of Sciences USA 99, pp 7610–7615

Fahey JW (2005) *Moringa oleifera*: a review of the medical evidence for its nutritional, therapeutic, and prophylactic properties. Part 1. Trees Life J 1(5):1–15

Faizi S, Siddiqui BS, Saleem R, Aftab K, Shaheen F, Gilani A-U-H (1998) Hypotensive constituents from the pods of Moringa oleifera. Planta Med 64:225–228

Farooq F, Rai M, Tiwari A, Khan AA, Farooq S (2012) Medicinal properties of Moringa oleifera: An overview of promising healer. J Med Plants Res 6 (27):4368–4374

Ghasi S, Nwobodo E, Ofili JO (2000) Hypocholesterolemic effects of crude extract of leaf of Moringa oleifera Lamin high-fat diet fed wistar rats. J Ethnopharmacol 69:21–25

Glover-Amengor M, Aryeetey R, Afari E, Nyarko A (2017) Micronutrient compositionand acceptability of Moringa oleifera leaf-fortified dishes by children in Ada-East district, Ghana. Food Sci Nutrit 5:317–323

Gopalakrishnan L, Doriya K, Kumar DS (2016) Moringa oleifera: a review on nutritive importance and its medicinal application. Food Sci Human Wellness 5 (2):49–56

Gupta S, Hemlata H, Tejavath K (2020) Synthesis, characterization and comparative anticancer potential of phytosynthesized mono and bimetallic nanoparticles using Moringa oleifera aqueous leaf extract. Beilstein Arch 1:95

Jain PG, Patil SD, Haswani NG, Girase MV, Surana SJ (2010) Hypolipidemic activity of Moringa oleifera Lam, Moringaceae, on high fat diet induced hyperlipidemia in albino rats. Revista Brasileira de Farmacognosia 20:969–973

Jaiswal D, Kumar Rai P, Kumar A, Mehta S, Watal G (2009) Effect of Moringa oleifera Lam. Leaves aqueous extract therapy on hyperglycemic rats. J Ethnopharmacol 123(3):392–396

Kushwaha S, Chawla P, Kochhar A (2014) Effect of supplementation of drumstick (Moringa oleifera) and amaranth (Amaranthus tricolor) leaves powder on antioxidant profile and oxidative status among post-menopausal women. J Food Sci Technol 51:3464–3469

Leone A, Spada A, Battezzati A, Schiraldi A, Aristil J, Bertoli S (2015) Cultivation, genetic, ethnopharmacology, phytochemistry and pharmacology of Moringa oleifera leaves: an overview. Int J Mol Sci 16:12791–12835

Ma ZF, Ahmad J, Zhang H, Khan I, Muhammad S (2020) Evaluation of phytochemical and medicinal properties of Moringa (Moringa oleifera) as a potential functional food. South African J Botany 129:40–46

Maldini M, Maksoud SA, Natella F, Montoro P, Petretto GL, Foddai M, De Nicola GR, Chessa M, Pintore G (2014) Moringa oleifera: study of phenolics and glucosinolates by mass spectrometry. J Mass Spectrom 49:900–910

Mehwish HM, Riaz Rajoka MS, Xiong Y, ZhengK, Xiao H, Anjin T,… & He Z (2020) Moringa oleifera–a functional food and its potential immunomodulatory effects. Food Rev Int, 1–20

Meireles D, Gomes J, Lopes L, Hinzmann M, Machado J (2020) A review of properties, nutritional and pharmaceutical applications of Moringa oleifera: integrative approach on conventional and traditional Asian medicine. Adv Tradit Med, 1–21

Mohammed S, Manan FA (2015) Analysis of total phenolics, tannins and flavonoids from Moringa oleifera seed extract. J Chem Pharmaceut Res 7:135–137

Moyo B, Oyedemi S, Masika PJ, Muchenje V (2012) Polyphenolic content and antioxidant properties of Moringa oleifera leaf extracts and enzymatic activity of liver from goats supplemented with Moringa oleifera leaves/sunflower seed cake. Meat Sci 91:441–447

Muangnoi C, Chingsuwanrote P, Praengamthanachoti P, Svasti S, Tuntipopipat S (2012) Moringa oleifera pod inhibits inflammatory mediator production by lipopolysaccharide-stimulated RAW 264.7 murine macrophage cell lines. Inflammation 35:445–455

Muhammad HI, Asmawi MZ, Khan NAK (2016) A review on promising phytochemical, nutritional and glycemic control studies on Moringa oleifera Lam in tropical and sub-tropical regions. Asian Pacific J Tropical Biomed 6:896–902

Okumu MO, Ochola FO, Mbaria JM, Kanja LW, Gakuya DW, Kinyua AW, Okumu PO, Kiama SG (2017) Mitigative effects of Moringa oleifera against liver injury induced by artesunate-amodiaquine antimalarial combination in wistar rats. Clinical Phytoscience 3:18

Padayachee B, Baijnath H (2020) An updated comprehensive review of the medicinal, phytochemical and pharmacological properties of Moringa oleifera. South African J Botany 129:304–316

Pari L, Kumar NA (2002) Hepatoprotective activity of Moringa oleifera on antitubercular drug-induced liver damage in rats. J Med Food 5:171–177

Sahakitpichan P, Mahidol C, Disadee W, Ruchirawat S, Kanchanapoom T (2011) Unusual glycosides of pyrrole alkaloid and 4′-hydroxyphenylethanamide from leaves of Moringa oleifera. Phytochemistry 72:791–795

Saini RK, Sivanesan I, Keum Y-S (2016) Phytochemicals of Moringa oleifera: a review of their nutritional, therapeutic and industrial significance. 3 Biotechnology 6:203

Sodvadiya M, Patel H, Mishra A, Nair S (2020) Emerging insights into anticancer chemopreventive activities of Nutraceutical Moringa oleifera: molecular mechanisms, signal transduction and in vivo efficacy. Curr Pharmacol Reports, 1–14

Sudha P, Asdaq SM, Dhamingi SS, Chandrakala GK (2010) Immunomodulatory activity of methanolic leaf extract of *Moringa oleifera* in animals. Indian J Physiol Pharmacol 54(2):133–140

Szlachetka K, Kut P, Stępień A (2020) Cytotoxic and anticancer activity of Moringa oleifera. Europ J Clin Exp Med 18(3):214–220

Talalay P, Talalay P (2001) The importance of using scientific principles in the development of medicinal agents from plants. Acad Med 76(3):238–247

Vats S, Gupta T (2017) Evaluation of bioactive compounds and antioxidant potential of hydroethanolic extract of Moringa oleifera Lam from Rajasthan, India. Physiol Molec Biol Plants 23:239–248

Verma AR, Vijayakumar M, Mathela CS, Rao CV (2009) In vitro and in vivo antioxidant properties of different fractions of Moringa oleifera leaves. Food Chem Toxicol 47:2196–2201

Waterman C, Rojas-Silva P, Tumer TB, Kuhn P, Richard AJ, Wicks S, Stephens JM, Wang Z, Mynatt R, Cefalu W, Raskin I (2015) Isothiocyanate-rich Moringa oleifera extract reduces weight gain, insulin resistance, and hepatic gluconeogenesis in mice. Molecular Nutr Food Res 59:1013–1024

William F, Lakshminarayanan S, Chegu H (1993) Effect of some Indian vegetables on the glucose and insulin response in diabetic subjects. Int J Food Sci Nutr 44 (5):191–196

Yan G, Liping S, Yongliang Z (2020) UPLC-Q-Orbitrap-MS2 analysis of Moringa oleifera leaf extract and its antioxidant, antibacterial and anti-inflammatory activities. Nat Prod Res 34(14):2090–2094

Zainab B, Ayaz Z, Alwahibi MS, Khan S, Rizwana H, Soliman DW, … Abbasi AM (2020) In-silico elucidation of Moringa oleifera phytochemicals against diabetes mellitus. Saudi J Biol Sci 27(9):2299–2307

Classical Genetics and Traditional Breeding in Moringa

5

R. Balakumbahan and N. Manikanda Boopathi

Abstract

Basic research advances in plant genetics and crop breeding methods have been translated into crop improvement and contributed to enhance food security and agricultural sustainability. Such impact has also been witnessed in Moringa and this chapter provides an account of cultivars that have been developed using different conventional breeding methods and strategies. Further, it also provides an overview of available Moringa ecotypes in India and the ways and means to explore these valuable genetic resources for further genetic improvement.

5.1 Introduction

Among the Indian originated crop, Moringa is reputedly referred to as a nutritious vegetable due to its rich content of vitamins, minerals, antioxidants and several other health promoting phytochemicals. As the slender, pendulous and thin-shaped immature fruits and curved pods of Moringa looks like a stick used for beating the drum, Moringa has been named as drumstick tree. Similarly, Moringa has also been called a radish tree as the pods look a lot like the siliqua of the radish. Moringa is also referred to in other names such as miracle tree, horseradish tree, mother's tree and West Indian Ben.

The botanical name of the Moringa plant which is widely cultivated in tropics and semitropical areas is *Moringa oleifera* Lam and it belongs to Moringaceae family. Moringaceae consists of only a single genus, *Moringa* (and hence also referred Monogeneric family). It has unique three-valved fruit which is slender, elongated and non-dehiscent pod and contains winged seeds. The leaves of *M. oleifera* are exclusively arranged in tripinnate pattern with 12–18 mm long leaflets. It has yellow or white petioles with or without red streaks.

The tree is native to northwest India, especially it is believed to be originated from foothills of north-western Himalayas, as there is booming wild ecotypes of Moringa in the sub-Himalayan tract (particularly in the riverside of Chenab that streams eastwards to the Sarda and in the Tarai zone, Uttar Pradesh, India). However, today Moringa is found worldwide and it is widespread in India, Philippines, Sri Lanka, Pakistan, Singapore, Thailand, Malaysia, Burma, Myanmar, West Indies, Egypt, Cuba, Jamaica and Nigeria (Ramachandran et al. 1980).

R. Balakumbahan
Department of Vegetable Crops, Horticultural College and Research Institute, Tamil Nadu Agricultural University (TNAU), Periyakulam, India

N. M. Boopathi (✉)
Department of Plant Biotechnology, Centre for Plant Molecular Biology & Biotechnology, TNAU, Coimbatore, India
e-mail: nmboopathi@tnau.ac.in

N. M. Boopathi et al. (eds.), *The Moringa Genome*, Compendium of Plant Genomes,
https://doi.org/10.1007/978-3-030-80956-0_5

A major economically important part of Moringa is the immature and tender pods as it has been consumed widely as nutritious vegetables especially in South India. Tender pods are cut into pieces and they are the important ingredient in culinary preparations and making of pickles. In several parts of the world, nowadays, Moringa leaves and flowers (besides green pods) are used to prepare nutritionally rich soup or taken as leafy vegetables.

Seeds are consumed after frying as that of peanuts. Ben or Behn oil is extracted from the Moringa seeds which is edible and has high economic value among the plant-derived oils. They are also used as lubricants, in the cosmetics industry and for illumination. The crushed and left-over seed kernels and cake after oil extraction has multi-fold applications: it can be used as fertilizer, fodder and water clarification agent. Recently, the Moringa biomass (leaves, petioles, young twigs) is also explored as a nutritionally rich fodder material. The Moringa gum, oozing out from the stem upon injury or physical damage, is having a good value in the European market and largely has its applications in calico printing as one of the important organic substances.

Irrespective of these multi-purpose uses and economic importance in the global market, only a few genetic improvement programs have been attempted to evolve novel cultivars and hence there are only a few commercially cultivated varieties (detailed below; Sect. 5.4).

5.2 Classical Genetics in Moringa

Among the 13 species of *Moringa* (see Chap. 1), only one species, viz. *M. oleifera* is extensively cultivated. However, vast morphological and genetic variability has been reported in this species. It is believed that such large genetic variation might be due to the cross-pollination that frequently occurs in Moringa. Detailed information on the extent of genetic variation present in *M. oleifera* is extremely essential to design a sound classical breeding and Moringa improvement program to improve the yield and quality of Moringa. A comprehensive report on the genetic diversity of different *Moringa oleifera* accessions has been provided in Chap. 7. Such studies provide information on genotypes of wide genetic origin that can be used as potential donor parent(s) in the upcoming Moringa breeding program.

Despite the above progress, a complete picture of morphological and genetic diversity patterns in the available Moringa germplasm and their cultivation requirement is really scarce. However, such information is required for effective genetic improvement program, domestication process and high-throughput cultivation. Studies on Moringa genetics also have several other major economic impacts such as increased yields, improved pest and disease resistance, herbicides resistance or to increase their nutritional value. Therefore, lack of information on the above classical genetics parameters is regarded as an impediment and it greatly affects the efficacious evolution of novel Moringa cultivars.

5.3 Traditional Breeding Strategies Used to Breed Novel Varieties in Moringa

There are two major approaches in plant breeding that lay the foundation for genetically improve the crop plants, including Moringa: conventional and molecular (or unconventional) breeding approaches.

Conventional breeding (also referred to as classical or traditional breeding) is evolving new varieties (or cultivars) of crop plants by using natural processes and older tools such as selection, hybridization, backcross and mutagenesis. For example, in hybridization and backcrossing programs, breeders eventually stack desirable traits from different but strongly associated ecotypes/landraces/wild plants into an elite cultivar using the techniques of crossing (hybridization) and backcrossing to the recurrent parent. Such a program leads to a reshuffling of the pre-existing gene or allele combinations present in the parents into a desirable set of combinations that improves the expression of target traits.

Thus, the plant genome is manipulated within the natural genetic limits of the species that are cross-compatible and such a conservative plant breeding approach has long (hundreds of years) been shown to be useful in evolving novel and desirable traits. However, those methods and techniques largely depend on the mode of reproduction of the plant species: self-pollinating, cross-pollinating or clonally propagated plants.

Moringa is a cross-pollinated crop and hence the following conventional methods are suitable for the development of a successful cultivar: mass selection, recurrent selection, family selection, hybridization and synthetics.

Selection is the oldest method of Moringa genetic improvement and has been efficiently exercised by differentiating the genetic variation found in the target traits. Though selection has been employed in Moringa genetic improvement for a long period of time (see below), it is achieved mainly through selecting the progenies that are visually appealed and possess higher economic yield (such as number and weight of the pods). So far, there is no report on selecting the progenies based on the superior nutritional or any biochemical properties.

The promising Moringa lines that are identified in one region or country are introduced into the new region or country and such breeding method is described in Introduction. Jaffna type of Moringa is one such introduced cultivar in Moringa (see below).

Hybridization is the next widely used classical breeding approach in Moringa genetic improvement. Hybridization or hybrid breeding is the controlled crossing of compatible parents to create novel and desirable genetic variation. This is mainly achieved by cross-pollination under a controlled environment: by establishing a hybridization block and effecting the pollination by artificial means through series of steps such as emasculating the recurrent parent, collecting the donor pollen and dusting on the emasculated recurrent parent, covering the pollinated ovary to avoid further pollination, etc., Though there are other unique hybridization schemes (such as diallele crosses) that have the opportunity to generate exclusive genetic combinations, they have been rarely explored in Moringa.

When the gene of interest is unavailable in nature and can be induced, a mutation breeding approach may be explored. An initial attempt in introducing genetic variation through chemical-induced mutagenesis at this institute (Udhayakumar et al. 2019) has illustrated the potential of this approach in Moringa but realizing its full potential in evolving new cultivar has yet to be demonstrated.

Additionally, if the useful trait(s) is reported in the landraces or wild relatives of the *Moringa*, such traits can be effectively transferred to cultivated Moringa species via a backcross breeding program. Though biotic and abiotic stress-resistant *Moringa* spp., are available in nature, the efficient introgression of these traits into cultivated Moringa is still considered as an unsuccessful process due to incompatibility between the wild and cultivated Moringa species.

The fundamental types of cultivars evolved through conventional breeding approaches are pure lines, open-pollinated, hybrid, clonal, apomictic and multi-lines. The basic principle in developing those cultivars is to maintain genetic purity and sustain the productivity under the natural mating system. The following sections provide promising ecotypes and cultivars that were evolved and utilized in commercial cultivation.

5.4 Moringa Ecotypes

Morphologically, cultivated *Moringa oleifera* is a medium-sized highly cross-pollinated fast-growing perennial tree indigenous to India. A wide range of morphogenetic diversity is seen in this subcontinent and spread to nearby countries like Sri Lanka (Kantharajah and Dodd, 1991; Olson 2002; Parrota 2009; Pandey et al. 2011; Leone et al. 2015). Until the dawn of the twenty-first century, Moringa was a homestead crop only (NRC 2006) and systemic large scale Moringa cultivation was slowly started in the southern part of India from the late 90 s (Verdcourt 1985; Lu and Olson 2001, Gandji et al. 2018).

Since it is a perennial crop and vegetatively propagated through stem cuttings, many ecotypes are available in India due to its adaptability to the given region (Ramachandran et al. 1980; Paliwal and Sharma 2011; Leone et al. 2015; Agoyi et al. 2015). Crop improvement in Moringa was initiated only during the recent past, after knowing the nutritional importance of this crop and its wider adaptability to diverse climatic conditions. The common ecotypes of perennial Moringa available for cultivation (Ramachandran et al. 1980; Kumar et al. 2014) are provided hereunder.

5.4.1 Jaffna Moringa

Yazphanam Moringa (popularly known as Jaffna Moringa) was introduced from Sri Lanka and it was widely cultivated in certain regions of Southern India. It produces long fruits (usually 60–90 cm in length and weighs 200–300 g) with soft flesh and good taste (Fig. 5.1). Jaffna Moringa starts to yield from the second year of planting ($\sim$40 pods in the second year) and from the third year onwards it may yield a maximum of 600 pods per tree. It is now being cultivated on large scales in Tirunelveli and Tuticorin districts of Tamil Nadu, India, for the preparation of value-added Moringa products that have rich export value.

5.4.2 Chavakacheri Moringa

Indeed, it is an ecotype of Jaffna Moringa that is cultivated in certain parts of Tamil Nadu, India. It has pods with a unique morphological feature which can produce a length of 90–120 cm (Fig. 5.2). It yields 500–600 pods/tree/year. However, due to the long size of pods, this ecotype of Moringa is highly damaged during transportation.

Fig. 5.1 Morphological features of pods produced by Jaffna Moringa

5.4.3 Chemmurungai Moringa

Chemmurungai is yet another heavy yielding ecotype of Jaffna Moringa and flowering occurs throughout the year and hence fruits can be harvested irrespective of the seasons. Ends (and/or in the middle) of the pods are red in colour and hence it got the name, Chemmurungai (in Tamil, the word chemmai denotes red). It is medium-sized tree, bearing long pods.

5.4.4 Moolanur Moringa

A unique perennial ecotype of Moringa that is largely grown in and around Moolanur, Karur, Dharapuram regions of Tamil Nadu, India is

Fig. 5.2 Morphological features of pods produced by Chavakacheri Moringa

Moolanur Moringa. It possesses medium-sized pod (with a length of 45–50 cm and weighs 120 g) and it yields generally 200 kg of pods per year. Usually, due to its perennial nature, the trees are left as such (i.e., without pruning) for a minimum of 15 years.

5.4.5 Valayapatti Moringa

Valayapatti Moringa is an alternative perennial Moringa ecotype that is widely cultivated in and around the Usilampatti and Andipatti zone of Tamil Nadu, India. It produces around 65 cm long pods (Fig. 5.3) that weigh about 120 g. The annual pod yield of this ecotype is relatively high compared to the other ecotypes (usually, 1000–1200 pods/tree per annum).

5.4.6 Kattumurungai

Moringa concanensis, which is being found in certain forest zones of Tamil Nadu, India, is

Fig. 5.3 Morphological features of pods produced by Valayapatti Moringa

referred to as Kattumurungai and the pods and leaves are used by the tribal people as an important vegetable or medicinal supplement to lactating mothers.

5.4.7 Kodikkal Moringa

As a support crop in the Betel Vine gardens of Tiruchirapalli district of Tamil Nadu, India, Kodikkal Moringa is being cultivated. Since the trees are having small structures and tiny leaves, they are mostly preferred for Betel Vine cultivation instead of other small tree species. Besides providing support to the vine growth, it also delivers short (20–25 cm long) and thick fleshed pods and leaves, which are found to be appetizing and delicious and frequently used for culinary purposes in these regions.

5.4.8 Palmurungai

It is commonly used for its unique taste due to its pods that are rich in thick pulp and grown in certain pockets of Kanyakumari districts Tamil Nadu, India.

5.4.9 Punamurungai

Though Punamurungai is not cultivated widely on commercial scale, it is found in the kitchen gardens of Tirunelveli and Kanyakumari districts of Tamil Nadu, India.

5.4.9.1 Palamedu Moringa

Another ecotype in Tamil Nadu, India is Palamedu Moringa which produces 60 cm long pods with 95–100 g weight and provides ~100 pods/tree in a year.

5.5 Cultivars Developed Using Conventional Breeding Methods

The main breeding method followed in Moringa is a clonal selection from the ecotypes available in the regions of Moringa cultivation. Primary systematic breeding works for Moringa crop improvement were initiated at Horticultural College and Research Institute, Tamil Nadu Agricultural University, Periyakulam, India during the 1980s. Consequently, two seed propagated Moringa varieties have been released from this institute for commercial cultivation (Ram et al. 2020).

5.5.1 PKM 1 (Periyakulam 1)

Moringa cultivar, PKM1 is a pure line selection from selfed progenies of the ecotype collected from Eppodumvendran region, Tirunelveli district, Tamil Nadu, India. The selection was advocated by retaining progenies that have high yield with desirable pod characteristics and those selected progenies were further put into generation advancement through selfing. The key characteristic of PKM1 is its 45–60 cm long pods with tasty flesh, which is highly preferred by the South Indians. PKM1 is propagated mainly through seeds and it continues to grow to a height of about 4–6 m at the end of the first year with 6–12 branches. It starts to yield six months after sowing and the peak pod harvest season is March to August.

PKM1 produces pinnate patterned compound leaves which may be in ~40 cm length and 20–30 cm width with small leaflets. The leaves are dark green on the top and pale or faded green on the bottom. Flowers are produced in clusters (each cluster may have 50–150 flowers), but one pod develops per cluster (however, 2 to 4 pods are also found in some cases). Pods can obtain a length of up to 70 cm with a girth of 6.3 cm that weighs ~150 g (Fig. 5.4). Both pods (which attain edible maturity 65 days after flowering) and the leaves of PKM1 are used as nutritive vegetables and greens.

On average, PKM1 annually yields 218 pods (which may weigh up to 35 kg) per tree and from one hectare it has been estimated that PKM1 yields 52.8 tonnes of pods in a year. Ratoon cropping in PKM1 is generally practiced for four years. However, the trees should be pruned at one meter from ground level once the peak harvest is completed in a year (Sauveur, 2001).

5.5.2 KM 1 (Kudumianmalai 1)

KM1 is also an annual Moringa developed through a pure line selection strategy. Similar to PKM1, it is also propagated by seeds and starts bearing from the sixth month onwards, and annually it provides 226–328 pods per plant. The fruits are 32–37 cm long and they can be easily harvested as the KM1 plants are shrubs in nature. After the first harvest, the plants pruned and allowed the main stem with a height of 1 m (above the ground) and grown as ratoon crop and ratoon cropping is followed for 2–3 years.

5.5.3 Dhanraj

KRC College of Agriculture, Arabhavi, Karnataka, India, has released an annual drumstick cultivar, viz., Dhanraj which is also propagated through seeds. It yields 150–200 pods per tree per year with a pod length of 35–40 cm.

Fig. 5.4 Morphological features of pods produced by PKM 1

5.5.4 Konkan Ruchira

Konkan Ruchira is a Selection from Vasai Local which was released by Dr. Balasaheb Sawant Konkan Krishi Vidyapeeth, Maharashtra, India. It is a medium-sized tree (up to 5.6 m in height) with dark green leaves. It produces medium-sized, dark green pods and is preferred as the best quality for culinary purposes. Flowering starts 90–100 days after planting and there will be 275 pods per tree per year.

5.5.5 Rohit

It is a perennial variety and released as pure line selection by evaluating 18 varieties for 5 years. Rohit starts to bear after 4–6 months of planting and continues to yield up to 10 years. Pods are 45–60 cm long, dark green and having soft and tasty pulp with good keeping quality or shelf-life.

5.5.6 Anupama

Anupama is a Moringa variety developed by Kerala Agricultural University, Trissur, India. It is an early bearer with an extended flowering period and high yielding capacity. The average yield per tree is 300 kg of pods. The pods are green in colour with average pod length of 55.5 cm.

5.5.7 Bhagya

Bhagya is an open-pollinated seedling selection from Dhanraj developed at University of Horticultural Science, Bagalkot, Karnataka, India. It is a medium-sized tree that grows to a height of 2–4 m and starts to flower 100–110 days after planting. The pods are green in colour with the fruit length of 60–70 cm. It yields 300–400 pods per tree per year.

5.5.8 PKM 2 (Periyakulam 2)

Horticultural College and Research Institute, TNAU, Periyakulam, India, released another cultivar with extremely long pod, PKM 2, which is a hybrid derivative of MP31 (Eppodumvendran Local) and MP28 (Arasaradi Local). Though, farmers practicing vegetative propagation, it can be effortlessly propagated through seeds and the ideal season for its cultivation in India is September–October. It is generally preferred to fit into different inter-cropping systems and home garden. The tree starts to yield in six months and the pods are extremely long (on an average, it produces 125.2 cm long with 8.3 cm girth, weighing 280 g pods; Fig. 5.5). Pods are rich in flesh, contains few seeds and tastier and more delicious. It produces 71.5 per cent higher pod yield over PKM 1 (see Sect. 5.5.1) and the average number of fruits produced by PKM 2 is 220 per tree per year which weighs 61.65 kg/tree

Fig. 5.5 Morphological features of pods produced by PKM 2

and under irrigated conditions it can yield up to 98 tonnes per hectare. PKM 2 can also be practiced as ratoon crop for 3 years and can be grown in most of the soil types (from sandy loam to clay loam) that have good drainage facilities.

5.6 Concluding Remarks

Moringa breeding programs conducted so far have greatly contributed in genetic improvement of major qualitative and quantitative traits in Moringa over the past couple of decades. However, by realising the stakeholder's preference and demands detected in the global Moringa market, it is exceedingly emphasized that the diversity of Moringa cultivars has to be increased in order to evolve novel cultivars that can adapt to wide range of soil and climatic conditions and with biotic and abiotic stress resistance besides having better quality characteristics. Hence, there is an enormous scope for Moringa breeding and it is hoped that advances in molecular breeding strategies (see Chaps. 9 and 10) would speed-up the process of releasing desirable cultivars in a short span of time.

References

Agoyi EE, Padonou EA, Amoussa W, Assogbadjo AE, Glèlè Kakaï R, Sinsin B (2015) Morphological variation, cultivation techniques and management practices of *Moringa oleifera* in SouthernBenin (West Africa). Int J Agrono Agricul Res 6(3):97–105

Gandji K, Chadare FJ, Idohou R, Salako VK, Assogbadjo AE, Glèlè Kakaï RL (2018) Status and utilisation of *Moringa oleifera* Lam: a review. Afr Crop Sci J 26(1):137–156

Kantharajah AS, Dodd WA (1991) Rapid clonalpropagation of *M. oleifera* Lam using tissue culture. South Ind Hort 39: 224–228

Kumar AR, Prabhu M, Ponnuswami V, Lakshmanan V, Nithyadevi A (2014) Scientific seed production techniques in moringa. Agric Rev 35:69–73

Leone A, Spada A, Battezzati A, Schiraldi A, Aristil J, Bertoli S (2015) Cultivation, genetic, ethnopharmacology, phytochemistry and pharmacology of *Moringa oleifera* leaves: an overview. Int J Mol Sci 16 (6):12791–12835

Lu L, Olson M (2001) Moringaceae. In: Wu Z-Y, Raven P (eds) Flora of China, Volume 8: Brassicaceae through Saxifragaceae. Missouri Botanical Garden Press, St. Louis, Missouri

National Research Council (NRC) (2006) Lost crops of Africa: Volume II: vegetables, development, security and cooperation. National Academy of Science, Washington, DC, pp 247–267.

Olson ME (2002) Combining data from DNA sequences and morphology for a phylogeny of Moringaceae (Brassicales). Syst Bot 27:55–73

Paliwal R, Sharma V (2011) A review onhorse radish tree (*Moringa oleifera*): a multipurpose tree with high economic and commercial importance. Asian J Biotechnol 3:317–328

Pandey A, Pradheep K, Gupta R, Nayar ER, Bhandari DC (2011) 'Drumstick tree' (*Moringa oleifera* Lam.): a multipurpose potential species in India. Genet Resour Crop Evol 58:456–460

Parrota JA (2009) *Moringa oleifera* Lam. Wiley-VCH Verlag GmbH and Co. KGaA, Weinheim

Patil V (2016) Drumstick, the game changer for him. The Hindu. April 3, 2016. http://www.thehindu.com/news/national/karnataka/drumstick-the-game-changer-for-him/article8427905.ece

Popoola JO, Obembe OO (2013) Local knowledge, use pattern and geographical distribution of *Moringa oleifera* Lam. (*Moringaceae*) in Nigeria. J Ethnopharmacol 150 (2):682–691

Ram HH, Kushwaha S, Dubey RK (2020) A glimpse of indigenous and minor vegetables of India. Indian Horticul 65(3):25–29

Ramachandran C, Peter KV, Gopalakrishnan PK (1980) Drumstick (Moringa oleifera): a multipurpose Indian vegetable. Econ Bot 34:276–283

Sauveur S A (2001) Moringa exploitation in the world: State of knowledge and challenges. Dar es Salaam. Tanzania. www.moringanews.org/actes/saintsauveuren.doc

Udhayakumar D, Paramaguru P, Boopathi NM, Swaminathan V (2019) Effect of gamma irradiation and ethyl methane sulphonate in annual moringa (Moringa oleifera L.) variety PKM-1. J Pharmacogn Phytochem 8(5), 2258–2261.

Verdcourt B (1985) A synopsis of the Moringaceae. Kew Bull 40:1–23

Cytogenetical Analysis of Moringa Genome

6

N. Manikanda Boopathi and M. Raveendran

Abstract

Advances in cytogenetics have resulted in the development of novel strategies in chromosomal banding patterns that allow the identification of individual chromosomes within a species. Differentiating chromosomes at the basic level is essential to draw essential genetic conclusions with respect to the given plant cells and such efforts have several folds of applications in cytological studies. Recent trends in structural, comparative and functional genomics experiments that supplement, accelerate or enhance the efficiency of Moringa-genome research towards product development are described here. This chapter also provides details on the current status of Moringa cytogenetics experiments and a way forward.

6.1 Importance of Cytogenetics in Crop Improvement

The study of DNA structure and organization that coiled or reduced to form chromosomes within the nucleus during the process of cell division is generally referred to as cytogenetics. The main aim of the cytological experiments is to identify the number and morphology of chromosomes by employing (i) chromosome banding techniques (classical cytogenetics) and (ii) hybridization techniques achieved by means of fluorescently labeled probes (molecular cytogenetics).

Except in the reproductive and liver cells of animals, all the cells of biological systems possess chromosomes with constant numbers and morphological features in the given species. Thus, the number and morphological features are used as an important descriptor of each species. For example, in Moringa, there are 28 (2n = 14) chromosomes and in humans, the chromosomes number are 46 (2n = 23).

In order to more precisely understand the evolution, genetics, genetic recombination and karyotypic stability of the investigated biological species, molecular cytogenetic tools have been employed as a key method in this era of genomics.

As efficient and robust methods of introducing novel variation in crop species, breeders always prefer cytogenetic manipulation of the given chromosome sets and evolve new crop cultivars. Besides, cytogenetics has also shown its utilities in crop improvement as detailed below:

1. Identifying and characterizing chromosome segments or chromosomes that are having various origins (which may be *intra-* and *inter*-specific).

N. M. Boopathi (✉) · M. Raveendran
Department of Plant Biotechnology, Center for Plant Molecular Biology and Biotechnology, Tamil Nadu Agricultural University, Coimbatore, India
e-mail: nmboopathi@tnau.ac.in

N. M. Boopathi et al. (eds.), *The Moringa Genome*, Compendium of Plant Genomes,
https://doi.org/10.1007/978-3-030-80956-0_6

2. Identification and validation of chromosomal markers linked with agronomically and economically important traits that allow plant breeders to efficiently introgress and track those traits in the progenies.
3. Finding the locations of genic and non-genic elements on the DNA sequences along the chromosomes and unravelling (in multidimensions) evolution and behaviour of physical chromosome upon addition and/or elimination of chromosomal segments and its impact on cell division and other cellular processes.
4. Cytogenetics has long been used as a key tool to document interspecific hybridization that helped to generate large numbers of commercially important variations in crop plants. Cross-pollination occurs in nature as well as in the controlled hybridization program, among the cultivated crop species and their wild relatives and exchange of chromosomal segments through homoeologous recombination is found widespread. Benavente et al. (2008) have shown that cytogenetic techniques were useful to identify the chromosomal segment from a donor species (i.e. the wild) that has been introgressed into a recipient species (i.e. the crop) by citing several hybrids and their derivatives as examples. As illustrated by them, all the cytogenetical tools (starting from the classical to the trendy molecular tools) can be used efficiently to distinguish the genomes and highlight the chromosomal regions that were exchanged or modified in due course in the investigated crop plants. They have also shown documental evidence that supported interspecific hybrids and fertility restoration by employing various strategies of cytogenetics.

6.1.1 Historical Perspectives of Cytogenetics in Crop Improvement

Revolutionary cellular research on maize (*Zea mays*) by Barbara McClintock has profoundly motivated crop cytogenetics. She undoubtedly laid the foundation for explicit documentation of each chromosome and led to develop several discoveries on the structure and frequent fluctuating behavior of the chromosomal rearrangement in the maize genome. The link between chromosome length and its arm ratio was deliberately established by her research finding that employed carmine-based chromatin staining procedures and she could professionally differentiate diverse chromosomes in a single meiotic nucleus. The same strategy was extended to other crops and cytogenetic maps were developed successfully in rice (*Oryza sativa*), sorghum (*Sorghum propinquum*) and tomato (*Lycopersicon esculentum*).

But later it was noticed that for the similar-sized chromosomes, more refined technological advancements are needed to improve the cytogenetic resolution and karyotyping (which is an artificial diagram of the genome in the form of chromosomes by pairing and ordering all the chromosomes of an organism). To this end, a fluorescent dye, quinacrine, was employed in 1968 at the Caspersson laboratory, to develop supplementary banding patterns on plant chromosomes and such bands were called as Q-bands. Q-bands provided additional resolution of alike sized chromosomes. Four years later, Giemsa C banding was developed by Vosa and Marchi and it was compared with Q-banding that was generated with the chromosomes of bean (*Viciafaba*), keeled garlic (*Allium carinatum*) and maize.

Such developments in chromosome-banding methods greatly advanced the practicality of utilizing the information on somatic chromosomes as it was found to be relatively simpler to obtain the information than meiotic chromosomes. Further advancement (such as cold pretreatments of the plant tissue samples) was also found to be useful for improved and better visualization of chromosomes when combined with a different modified procedure that employs various concentrations of carmine that were optimized for several economically important crops. Despite the considerable progress in chromosome characterization, development of in situ hybridization (ISH) techniques has made a quantum leap in cytogenetics as it allowed superior imaging of specific DNA sequences on chromosomes.

ISH utilized radioactive tracer or modified nucleotides (a chemical moiety (such as biotin, digoxigenin, or fluorescent moieties) that is attached to the nucleotide) which binds to the complementary sequences in cells/nuclei and/or on individual chromosomes. Such binding facilitates easy visualization of the ISH probes and their localization on the chromosomes.

Langer-Safer et al. (1982) have first reported the Fluorescence in situ hybridization (FISH) and since then it has been largely applied in several crops and of late, after the attachment of CCD camera and imaging process of FISH signals, its applications in cytogenetical studies are widespread. In addition to that, there are continuous advancements in FISH and it led to the introduction of several novel improvements such as multi-colour FISH (McFISH), genomic in situ hybridization (GISH), chromosome image analyzing system (CHIAS), flow cytometry, pulse-field gel electrophoresis (PFGE), chromosomal microdissection and microcloning. Such advancements have greatly increased the visualization of not only the individual chromosomes with high resolution but also efficient physical localization of target DNA fragments on individual chromosomes with high precision (Gupta and Dhar 2004). Besides, FISH has also been extensively employed in mapping repetitive DNA sequences, low-copy-number sequences, chromosome recognition in a cocktail and establishing genomic relationships among chromosomes in the polyploid organisms or ecotypes with high similarity.

The key application of today's cytogenetic experiments is precisely identifying the locus and order of two or more DNA fragments (usually referred to as molecular markers) on the given chromosome. The capacity to discriminate the positions of two loci as proximal or distal to the centromere is generally defined as axial resolution and it is indicated in either base pairs or spatial unit (centi McClintock, cMC; (Figueroa and Bass 2010). Several studies have reported axial resolutions ranging from <0.7 kb in fiber-FISH to 10,000 kb in some metaphase(M)-FISH and helped to make clear differences among the DNA fragments on the chromosomes and it has been predicted that an axial resolution limit of ~600–700 bp can be obtained with the current epifluorescence microscopy (Figueroa and Bass 2010). Such efforts are particularly useful in high-density genetic map construction using molecular markers (Boopathi 2020).

FISH has been used to unravel the unearthed information on individual chromosomes such as their structure, mutation and evolution and differentiating closely related species based on their chromosome organization, besides characterizing each chromosome by localizing satellites, chromosome arms and centromeres. These studies would have implications in designing novel probes for RFLPs and mapping of agronomically important traits and characterizing plant genomes for gene duplication, chromosome synteny and evolution (Gupta and Dhar 2004). Further, FISH mapping has also been useful in the construction of physical maps and joining the linkage maps and physical maps.

The most substantial recent advance in FISH, in the above lines, is the use of synthetic oligonucleotides (oligos) to prepare probes. Oligos can be artificially synthesized to mimic repetitive DNA sequence or a specific chromosomal region or even an entire chromosome and labeled with fluorescent dyes. Such probes can be used for extensive comparative cytogenetic mapping among genetically related species (Jiang 2019) and their genetic improvement by translating the genetic information identified in one species to the other.

Thus, FISH has shown its widespread applications in structural, comparative and functional genomics and accelerated the crop-genome research towards efficient improvement of crop plants in this genomics era. Further involvement of other advanced strategies such as molecular and immune-chromatin cytology (for example, combining FISH, chromatin immunoprecipitation and immuno-cytochemistry), novel chromosome preparation methods and high-resolution imaging would offer novel insights in chromosome organization at several dimensions.

6.2 Cytogenetics of Moringa Genome

Initial cytological studies have discovered that *Moringa oleifera* Lam., possess 2c genome size of 1.2 pg (Ohri and Kumar 1986) and its diploid number (2 N) was 28 (Mendior et al. 2004). Totally, 14 bivalents in diakinesis were found as the meiotic behaviour of *M. oleifera*. Further studies have revealed that there were more than 88% pollen viability and rare meiotic abnormalities in which the most affected cell cycle was metaphase I. It was additionally established that there was a development of unreduced gametes due to tripolar spindles in metaphase II (Silva et al. 2011).

In another study (Anwar 2016), cytological features (such as cell shape and size, mitotic chromosomes and nucleolus appearance) of *M. oleifera* and *M. stenopetala* were explored and he reported that active meristematic cells could be categorized into three types: Type I (including dividing cells), Type II (meristematic cells with diffused chromatin) and Type III (cylindric cells with diffused chromatin) and further reported the percentages of dividing cells as the mitotic index.

It was also evident from that study that mitotic index was considerably high at cells of *M. oleifera* than that of *M. stenopetala*. The scored diploid number (2 N) of metaphase plates of the plants of the two Moringa species was mostly 28 and they were small-sized chromosomal sets. However, large nucleolus at metaphase and even at anaphase or telophase appeared only with *M. stenopetala* whereas no nucleolus body was detected in mitotic cells of *M. oleifera* except at interphase and in some cases at prophases (Anwar 2016). Nucleolus body appeared in metaphase and anaphase/telophase cells of *M. stenopetala* may be due to the species-specific difference and it may be used as a cytological marker to distinguish the plants of these two Moringa congeneric species.

6.3 Way Forward

In scientific history, a handful number of techniques have dominated the research field for more than four decades and FISH is one such technique that has been successfully used in cytogenetic research since the 1980s. On the other hand, due to the unavailability of robust DNA probes in many crops, especially in non-model crops, the application of FISH techniques has been limited down since the dawn of the twentieth century. To alleviate this limitation, synthetic oligo probes were introduced. In addition, several exciting new developments were also introduced recently. For example, exploration of CASFISH (employing fluorescently labelled nuclease-deficientCas9 (dCas9) proteins to detect chromosomal regions without employing DNA denaturing process (otherwise it would interrupt the organization of genome in the fixed or living cells) and single-molecule RNA-FISH (smRNA-FISH) techniques (employed to quantify the transcription of multiple genes or non-coding RNAs within single cells) would advance high-end FISH applications in plants.

With respect to cytogenetics studies in Moringa, a lot more has to be done starting from karyotyping to merging of physical mapping with chromosomal mapping. It is sincerely hoped that advances in molecular cytogenetics tools and strategies, would pave new ways and means to fill these cytogenetics research gaps in Moringa. Although not yet reported in *M. oleifera*, cytogenetics-based manipulation of Moringa chromosomes can be used to induce polyploidy as it has more impact on the improvement of nutritional and the medicinal properties of this species. Moringa molecular cytogenetics can also be combined with genomics, reverse genetics and comparative genomics strategies which will eventually lead to the discovery and isolation of novel genes (that are agronomically, economically and nutritionally important). Such cytogenetic information will be extensively utilized in precision breeding strategies that help to genetically improve Moringa. Identification of such novel genes is not possible with the conventional Mendelian genetic analysis.

References

Anwar GM (2016) Studies of some cytological features on two Moringa Species (M. Oleifera And M. Stenopetala) Cultivated In Egypt. Minia J Agric Res Develop 36(4):601–611

Benavente E, Cifuentes M, Dusautoir JC, David J (2008) The use of cytogenetic tools for studies in the crop-to-wild gene transfer scenario. Cytogenetic Genome Res 120(3–4):384–394

Boopathi NM (2020) Linkage map construction. In: Genetic mapping and marker assisted selection, pp 179–227. Springer, Singapore

Figueroa DM, Bass HW (2010) A historical and modern perspective on plant cytogenetics. Briefings Funct Genomics 9(2):94–102

Gupta PK, Dhar MK (2004) Advances in molecular cytogenetics: potential for crop improvement. In: Plant Breeding, pp 97–114. Springer, Dordrecht

Jiang J (2019) Fluorescence in situ hybridization in plants: recent developments and future applications. Chromosome Res 27(3):143–164

Langer-Safer PR, Levine M, Ward DC (1982) Immunologicalmethod for mapping genes on Drosophila polytene chromosomes. Proc Natl Acad Sci U S A Biol Sci 79:4381–4384

Mendior MS, Diaz MGQ, Alcantara MTB, Hilario OJ, Mateo P, Maghirang RDM (2004) Cytological Studies of selected medicinal plants: Euphorbia pulcherrimaWilld. ex Klotz., Moringa oleifera Lam., Catharanthusroseus (L.) Don.,and Chrysanthemum indicum Linn. Philippine J Sci 134(1):31–37

Ohri D, Kumar A (1986) Nuclear DNA amounts in some tropical hardwoods. Caryologia 39:303–307

Silva N, Mendes-Bonato AB, Sales JGC, Pagliarini MS (2011) Meiotic behavior and pollen viability in Moringa oleifera (Moringaceae) cultivated in southern Brazil. Genet Mol Res 10(3):1728–1732

Genetic Diversity of Moringa (*Moringa Oleifera* Lam.)

7

T. N. Lakshmidevamma, J. Ugalat, K. A. Apoorva, S. P. Gautham Suresh, M. Doddamani, S. Kadam, R. S. Nayana, R. C. Jagadeesha, and B. Fakrudin

Abstract

Moringa (*Moringa oleifera* Lam.) is a widespread multipurpose tree with great potential as a high-value crop for its nutritive, therapeutic and prophylactic properties with several industrial applications. Almost every part of the tree is useful with varied end uses. It is an interesting plant owed to its bioactive compounds. *Moringa* species have been widely spread across the tropical and subtropical regions of the world and a total of 13 species of the genus *Moringa* are known. Moringa, being a cross-pollinated tree, high heterogeneity in many character forms is observed resulting in vast genetic diversity in the natural and cultivated accessions and substantial variation in quantitatively inherited traits has been documented which needs to be exploited for concentrated research towards its crop improvement. In-depth knowledge and understanding of the gene flow pattern and population genetic structure in moringa through molecular genetic diversity and population structure of worldwide collections is of great promise. In spite of its varied uses and morphological diversification, the number of accessions collected as germplasm and their conservation in gene banks is very meagre. Documentation of genetic diversity and conservation of germplasm is a necessity for strategic research as well as breeding programmes to develop elite varieties. This chapter highlights the genetic diversity in moringa, its significance in the contemporary nutritional security scenario and its exploitation for future programs.

7.1 Introduction

Moringa (*Moringa oleifera* Lam.) is a widespread multipurpose tree known for its nutritive, therapeutic and prophylactic properties with numerous industrial applications. It is an ancient seldom utilized tree but gaining importance recently due to its exceptional uses. It is being studied for nutritional attributes and as a livestock fodder crop. The fast-growing ability of the crop helps to withstand both severe drought and mild frost conditions and has made moringa, a hardy tree crop cultivated wildly across tropics and subtropics (Csurhes and Navie 2016). Globally, moringa is being promoted as a nutritious tree crop in marginal areas with less rainfall as well as homestead gardens (FAO 2020). This

T. N. Lakshmidevamma · J. Ugalat · K. A. Apoorva · S. P. G. Suresh · M. Doddamani · S. Kadam · R. S. Nayana · R. C. Jagadeesha · B. Fakrudin (✉)
Department of Biotechnology and Crop Improvement, College of Horticulture, University of Horticultural Sciences, Bengaluru campus, Bengaluru 560065, India
e-mail: bfakrudin@uhsbagalkot.edu.in

N. M. Boopathi et al. (eds.), *The Moringa Genome*, Compendium of Plant Genomes,
https://doi.org/10.1007/978-3-030-80956-0_7

tree has the prospective to increase nutrition, food security and boost rural development (Hsu et al. 2006). The long, slender, triangular seed-pod properties earned it the vernacular name 'drumstick tree' and from the horseradish taste of the roots 'horseradish tree' besides, ben oil tree or benzolive tree (Paliwal et al. 2011). It is popularly called as 'mother's best friend' in the Philippines, as it is consumed to increase a woman's milk production as well as for the treatment of anaemia (Estrella et al. 2000; Siddhuraju and Becker 2003). Recently this species has been dubbed as 'miracle tree' or 'natural gift' or 'mother's best friend' (Leone et al. 2015; Pirro et al. 2019).

Almost every part of the tree is useful either for its nutritional value or for commercial purposes, hence, it is also known as the 'natural nutrition of the tropics'. The leaves, flowers, fruits and immature pods are highly nutritious and consumed in Asian, African and American countries such as India, Pakistan, Sri Lanka, Burma, Indonesia, Philippines, Hawaii (Anwar and Bhanger 2003). The extracts from moringa leaves are used to treat malnutrition, augment breast milk in lactating mothers as they are rich in minerals, vitamins and other essential phytochemicals. It has the potential to use as an antioxidant, anticancer, anti-inflammatory, antidiabetic and antimicrobial agent. Moringa seed is extensively used in the purification of drinking water as it acts as a natural coagulant (Gopalakrishnan et al. 2016; Falowo et al. 2018). In various forms, this tree is used to treat turbid water in many countries (Suarez et al. 2003; Bhatia et al. 2007). In addition to food uses, moringa is being used for animal feed (Sánchez et al. 2006). When supplemented to the diet of dairy animals, moringa leaves improve dry matter intake, digestion and milk production, but without affecting the smell, taste or colour of milk (Sánchez et al. 2006). Further, leaf and seed powders are a good source of phytomedical compounds (Anwar et al. 2007). It is also a promising source for bioenergy (Adu-Dapaah et al. 2015).

7.2 Origin and Distribution

Moringa species have been widely spread across the tropical and subtropical countries of the world (Csurhes and Navie 2016). Moringa is thought to originate from northern India, specifically around Agra and Oudh and South of the Himalayan Mountains (Vogt 1996). It spread to eastern Africa during the beginning of the twentieth century (Mallenakuppe et al. 2015). The species is either grown commercially or present in wild across Asian, African, North American, Central American, the Caribbean, South American and Oceania counties (Acevedo-Rodríguez and Strong 2012). In recent times, Moringa is being grown throughout the Middle East, almost the whole tropical belt and is expanding to a diverse environment of tropical and subtropical regions in Africa and America (Pandey et al. 2011; Chang et al. 2019).

7.3 Genetic Resources of Moringa

Moringa belongs to the order Brassicales and the monogeneric family Moringaceae, which consists of about 33 species (Arora et al. 2013). Among these, 13 species, viz., *M. arborea, M. longituba, M. pygmaea, M. concanensis, M. borziana, M. hildebrandtii, M. stenopetala M. oleifera, M. ovalifolia, M. peregrina, M. rivae, M. drouhardi, M. ruspoliana* are found worldwide (Stephenson and Fahey 2004). Origin of moringa species could be traced back to different countries: *M. arborea*, native to Kenya; *M. borziana* from Somalia and Kenya; *M. rivae and M. stenopetala* originated from Kenya and Ethiopia; *M. pygmaea* native to Somalia; *M. longituba* from Ethiopia and Somalia; *M ruspoliana* native to Ethiopia; *M. ovalifolia* from Namibia and Angola; *M. drouhardii* and *M. hildebrandi* from Madagascar; *M. peregrine* from Red Sea and Horn of Africa; *M. concanensis* and *Moringa oleifera* indigenous to sub-Himalayan tracts of northern India, among which *Moringa oleifera* has been the most well-studied and used species for human consumption (Paliwal et al. 2011). Further, *M. arborea, M. ruspoliana, M. longituba, M. stenopetala M. rivae,* and *M. borziana* are in danger of extinction (Stephenson and Fahey 2004). Only *M. oleifera* is cultivated (Sánchez et al. 2006). Moringa cytological studies revealed that *Moringa oleifera* has 2c genome size of 1.2 pico gram (pg) (Ohri and Kumar 1986). The diploid chromosome number of this species is 2n = 28 (PROTA 2017).

Distribution of several ecotypes can be seen across India with several vernacular names, namely; Jaffna and Chavakacheri murungai (soft and taste fruits), Chemmurungai (red-tipped fruits), Kadumurungai with small and inferior fruits, Palmurungai and Punamurungai having a bitter taste, Kodikalmurungai with short fruit, wild Kadumurunga and Kodikkal Murungai (Kumar Ganesan et al. 2014). Two varieties viz., PKM-1; PKM-2 from Tamil Nadu and one variety, Bhagya from Karnataka, India, have been developed higher pod yield, pod quality and nutritional parameters. Several moringa varieties with desirable attributes also developed from Kerala, India (Kumar Ganesan et al. 2014). Even though vast variability is present as cultivated accessions and natural accessions, systematic studies towards the collection of wild and local forms as well as cultivated accessions across the globe and genetic diversity quantification are meagre. In spite of its varied uses and morphological diversification, the number of accessions collected as germplasm and their conservation in germplasm banks are just emerging across the world. The gap between the inherent genetic variability available and its poor representation in gene banks is a hindrance for crop improvement programs that needs to be addressed. So, gene/germplasm banks covering the entire genetic variability in Moringa is a necessity for concentrated research as well as breeding programmes to develop elite varieties.

Most of the genetic variation of *M. oleifera* is reported in India (Kumar Ganesan et al. 2014; PROTA 2017). The species has various cultivars, including some cultivated as annuals in temperate climates. DNA barcode details are available at the Barcode of Life Data Systems (Godino et al. 2015). Research centers concentrated on *Moringa oleifera* improvement across the globe are AVRDC (Taiwan), Moringa Philippines Foundation (Philippines), Moringa community (Zambia) and Rural development initiative (Zambia). Germplasm collections are stored at various facilities worldwide (Patricio and Palada 2015). The World Vegetable Center at Taiwan maintains collections of four moringa species, the majority of which are *M. oleifera* (Palada et al. 2017). Further, moringa collections are also maintained at the International Moringa Germplasm Collection in Mexico and Burkina Faso and the Philippines (IMGC 2017; PROTA 2017). The International Moringa Germplasm Collection houses living material of 12 of the 13 *Moringa* species at the Instituto de Biología, Universidad Nacional Autónoma de México (UNAM), Mexico (PROTA 2017).

7.4 Genetic Diversity in *Moringa Oleifera*

Genetic diversity is an essential pre-requisite in crop improvement. The chances of obtaining desirable transgressive segregants depend on the extent of parental genetic diversity: the greater the genetic diversity, the higher the prospects of desirable segregants. Diverse agro-ecological conditions, migration of genetic material due to genetic drift, gene flow, introduction/exchange of genetic stocks at national and international levels, along with adoptive as well as intensive artificial selection are responsible for the diversification of drumstick plants.

Moringa, being an outbreeding tree, high heterogeneity with high diversification in many characters are observed. The crop shows extensive diversity of morphological and biochemical properties, which serve as a resource for its genetic improvement (Godino et al. 2015). Wide variability was revealed like tree habit from deciduous to evergreen; shape of the tree varying from semi spreading to upright; flowering time wherein some trees flower throughout the year and some flower in two distinct seasons

(Ramachandran et al. 1980; Raja et al. 2013) and resistance to hairy caterpillar (Mgendi et al. 2010; Raja et al. 2013).

Thirty-six genetically diverse genotypes of drumstick were clustered into five clusters based on the content of vitamin C, protein, nitrogen, phosphorous, potassium, calcium, iron and magnesium. The analysis used measures the forces of differentiation at two levels, namely intra-cluster and inter-cluster levels, and thus help in the selection of genetically divergent parents for exploitation in hybridization programmes (Tak and Maurya 2015). Genetic diversity, population structure and correlation in Indian populations of *Moringa oleifera* studied by Rajalakshmi et al. (2019) revealed the high variability existent at an intra-population level.

7.5 Nutritional Diversity in Moringa

Moringa is one of the most useful trees that can be used for food, nutraceutical value and industrial purposes (Khalafalla et al. 2010). People use its leaves, flowers and fresh pods as vegetables, while others use it as livestock feed (Anjorin et al. 2010). The significance of moringa to address nutritional deficiencies is based on the abundance of vitamins, minerals and protein found in the leaves and pods. These include vital nutrients such as beta-carotene, iron, zinc, vitamin C and all the essential amino acids (Moyo et al. 2011). The plant is rich in other bioactive and anti-inflammatory compounds like phenolics and isothiocyanates, while being relatively low in anti-nutrients (Falowo et al. 2018; Waterman et al. 2014). Different parts of the plant are differing with respect to their nutritional value. 100 g of pod yields 26 cal of energy; 2.5 g protein; 0.1 g fat; 3.7 g carbohydrates; 4.8 g fibre; 0.05 mg vitamin B1; 0.07 mg vitamin B2; 0.2 mg vitamin B3; 120 mg vitamin C; 30 mg calcium; 24 mg magnesium; 110 mg phosphorous; 259 mg potassium; 3.1 mg copper; 5.3 mg iron and 137 mg sulphur. Whereas, fresh leaves yield about 92 cal of energy and 6.7 g protein; 1.7 g fat; 12.5 g carbohydrates; 0.9 g fibre; 0.06 mg vitamin B1; 0.05 mg vitamin B2; 0.8 mg vitamin B3; 220 mg vitamin C; 448 mg vitamin A; 440 mg calcium; 42 mg magnesium; 70 mg phosphorous; 259 mg potassium; 0.07 mg copper and 0.85 mg iron (Gopalakrishnan et al. 2016). The leaves are considered as a better source of protein, Ca and Fe while the pods are rich in P, Na and Mg. Both the edible parts (leaves and pods) more specifically leaves have the potential to be used as micronutrient supplements for food products (Mallenakuppe et al. 2015).

The presence of wide nutritional variation in Moringa could be accredited to varying genetic backgrounds of the plant, in terms of ecotype, cultivar as well as environmental factors (Sánchez-Machado et al. 2010). Nutritional composition of four species; *M. oleifera, M. peregrine, M. stenopetala, M. drouhardii* indicated high overall nutritive value, antioxidants and glucosinolates, with low oxalate content, whereas *M. oleifera* contained the highest amounts of β-carotene, ascorbate, α-tocopherol, iron and protein content and *M. peregrina* was the uppermost for antioxidants (Yang et al. 2006). Substantial variation for chlorophyll, protein, macronutrient and micronutrient content in the leaves of different genotypes of *M. oleifera* and *M. peregrina* was revealed, which indicated higher chlorophyll in *M. peregrine* and higher protein content in *M. oleifera* (19.1–32.9%) than *M. peregrine* (12.5–24.6%).

Similarly, Fe and Zn contents varied between 250–490 ppm and 26.7–58.3 ppm respectively in *M. oleifera* versus 285–403 ppm and 15.7–38.9 ppm in *M. peregrine*. Mp genotypes were mostly intermediate in K content compared to Mo genotypes (1.3–2.0%). Calcium content in both species ranged from 0.03 to 1.3%. The highest Mg content was found in Mp genotypes (1.0%). Genotypes also varied in their content in Zn, Cu, and Mn for both species (Hassanein 2018). A similar trend of macronutrients was reported on *M. oleifera* genotypes grown in the Philippines (Magat et al. 2009). Further in Ghana, different genotypes of *M. oleifera* varied in protein and Ca contents with a wider range of protein, 26–27% in Ghana and significantly differed in their contents of all studied nutrients (Asante et al. 2014). Saudi Arabia *M. oleifera* genotypes were reported to be more diverse than those of other regions for their nutrient contents (Nouman et al. 2016).

Various accessions of moringa from different agro-ecological regions of Rajasthan, Uttarakhand, Uttar Pradesh, Madhya Pradesh and Goa have revealed high variability in the quantity of phytochemicals (Nouman et al. 2016). Two cultivars from the two ecological regions Chapai Nawabganj (L1) and Pabna (L2) of Bangladesh indicated a wide variation in nutritional value for protein and mineral nutrients viz., Ca, Fe, P and K. Ten moringa accessions, selected from a survey of 60 moringa accessions revealed variation for nutrient contents (Islam et al. 2020). Polyphenolics composition, antioxidant activity and contents of selected nutrients in the leaves from seven cultivars of *M. oleifera* ('Tumu', 'Sunyaw', 'Kumasi', 'Techiman', 'China', 'Pakistan Black' and 'Pakistan White') in Pakistan revealed variation among cultivars (Nouman et al. 2016).

7.6 Polymorphism and Allelic Frequency

Molecular diversity among seven populations of *M. oleifera* from Africa was explored and molecular variance of 59.15% between populations was inferred. When compared with the Indian population, African population showed 18.59% of variation existed between the two populations (Muluvi et al. 1999). These seven populations also revealed a total of 236 amplifications, of which 157 (66.5%) were found to be polymorphic between or within the populations (Muluvi et al. 1999). The microsatellite marker analysis of 24 germplasm of moringa exhibited an average of two to six alleles per locus. The expected and observed heterozygosity varied from 0.361 to 0.761 and from 0.01 to 0.875, respectively (Wu et al. 2010). Taxonomic analysis of 75 accessions of Moringa from Nigeria revealed a high level of polymorphism between the samples from south-central and northern parts of the country (Abubakar et al. 2011).

The high degree of polymorphism (74%) with respect to the genetic relationship was reported between the accessions from Brazil (da Silva et al. 2012). 161 accessions of *M. oleifera* from Asia, Africa, South and North America, and the Caribbean revealed allelic diversity of 8.3 alleles per SSR marker with a total of 158 alleles in 131 wild accessions collected from Pakistan and 30 accessions obtained from Florida (Shahzad et al. 2013).

Further, the genetic diversity in these 12 Indian populations using SSR primers elucidated a total of 35 bands of which 29 bands exhibited polymorphism (82.86%) with 1.84 bands /primer. Analysis of molecular variance (AMOVA) revealed 2% diversity which could be attributed to differences among regions, 3% at the level of populations and 95% contributed by the genotypes within the populations (Kumar Ganesan et al. 2014). A total of 74 alleles with a range of 4 to 15 were detected in 70 accessions of *M. oleifera* with an average of 7.4 alleles per locus. Allele frequency varied from 0.214 to 0.671 with a mean of 0.477; gene diversity from 0.487 to 0.885 with a mean of 0.669 while the average PIC value was 0.633 (Popoola et al. 2017). The genotypes collected from different countries revealed the presence of a higher genetic variance of 1.80 and 0.13 for the Malaysian population and 0.30 and 0.19 for the international population, respectively (Rufai et al. 2013).

Three molecular marker techniques, *i.e.* random amplified polymorphic DNA (RAPD), inter simple sequence repeat (ISSR) and cytochrome P450 gene-based markers in eight Indian cultivars of *M. oleifera*, revealed 48.68, 48.57 and 40.00% polymorphisms, respectively (Saini et al. 2013). Genetic variability analysis in 97 *M. oleifera* accessions indicated an average of 59.6% polymorphism through ISSR and 70.13% with SRAP markers. The mean gene

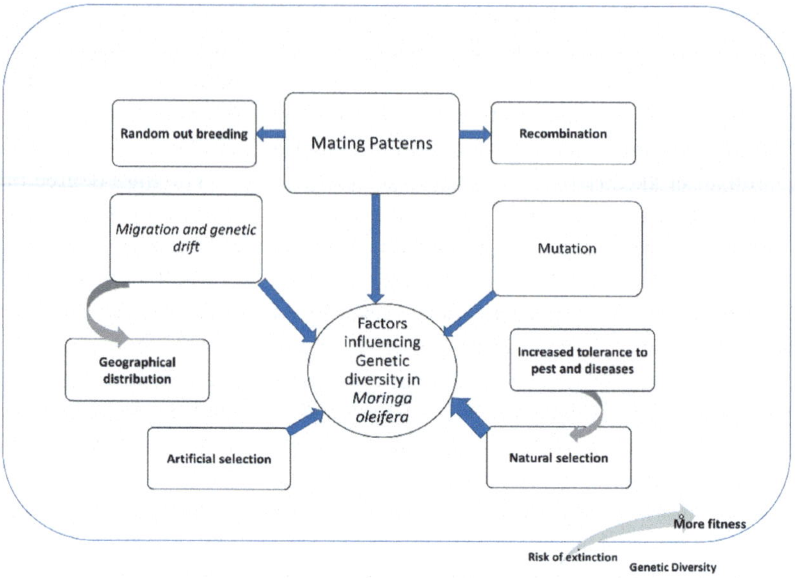

Fig. 7.1 Factors influencing genetic diversity in *Moringa oleifera*

diversity h (0.2), gene flow N_m (2.86) and a PhiPT value (0.14) estimated for the markers indicate the high variability existent at an intra-population level. Population structure results and cluster tree divided the samples into two groups: One group consists of Tamil Nadu accessions and the other comprises Andhra Pradesh and Odisha accessions (Rajalakshmi et al 2019). Figure 7.1 describes different factors that influence the Moringa genetic diversity.

Figure 7.1 Multiple factors affecting genetic diversity in a population of drumstick: Being a cross-pollinated crop, the genetic variability within and among the populations of moringa is derived from a wide assortment of genes and alleles which is affected by various factors. Mating patterns such as non-random outbreeding is common in cross-pollinated species. Random forces, such as population bottleneck causes changes in allelic frequency, which leads to genetic drift. Individuals within a population are subjected to spontaneous mutation, which evolves the species. Movement of individuals into or out of a defined location causes the migration of a set of alleles within a population. Natural disasters and competition for food and other resources favour certain individuals, naturally selecting them for further generations.

7.7 Principal Coordinate Analysis

Principal coordinate analysis (PCA) clusters and relates individuals or populations based on genetic distances. PCA of 20 genotypes of *M. oleifera* from different countries classified the twenty genotypes of *Moringa* into five major groups; two genotypes from Malaysia in one group. Dimension one of the PCA ranged from 0.49 to 0.92, while dimension two varied from −0.65 to 0.26 and dimension three varied from −0.65 to 0.68 (Rufai et al. 2013). PCA of 12 Indian populations of *M. oleifera* revealed very high genetic diversity in the Indian drumstick collection (Kumar Ganesan et al. 2014).

Phenotypic intraspecific variations among 40 accessions of *M. oleifera* collected from different agro-ecological zones of Nigeria revealed a high degree of intraspecific variability for reproductive characters. The first five principal component axes explained 61.40% of the overall variation with PC1 (23.92%) and PC2 (14.19%) contributing 38.11% of the total variation (Popoola et al. 2016). Commercially grown *Moringa oleifera* cultivars from India resulted in PCA plots where all the cultivars were assembled into four major clusters and genotypes belonging to a particular sub-cluster were grouped together (Saini et al. 2013).

7.8 Phylogenetic and Population Structure

In-depth knowledge and understanding of the detailed gene flow pattern and population genetic structure in Moringa is very meagre (Muluvi et al. 2004). Thus, the development of efficient molecular markers for *M. oleifera* is needed. The genetic diversity and population structure of worldwide collections of *M. oleifera* were investigated using 19 simple sequence repeat (SSR) markers along with a partial sequence of the chloroplast gene *atpB* which demonstrated large genetic diversity present in wild collections from Pakistan (Shahzad et al. 2013).

Parsimony analyses to infer the phylogenetic relationships of all 13 species of the genus *Moringa* revealed the four bottle trees in a basal paraphyletic assemblage, with the three species of slender trees, including the economically important *M. oleifera* forming a clade that is sister to a clade of the six species of tuberous shrubs and trees of northeast Africa (Olson 2002). 20 M. *oleifera* populations from Nigeria, using amplified fragment length polymorphism (AFLP) primer pairs clustered accessions along eco-geographical lines while others grouped disparately (Popoola et al. 2019).

Model-based population structure of 12 Indian *M. oleifera* populations using *K* values from 1 to 20 resulted in five clear populations instead of 12 natural populations. Cluster analysis and structure-based population study showed that no geographical isolation exists between genotypes collected from the southern and northern parts of India (Kumar Ganesan et al. 2014). Population structure analysis of 161 M. *oleifera* accessions collected from nine other countries of the world was grouped into three clusters and there was a sharp decrease in delta-K values from K2 to K3, with a plateau at K4 (Shahzad et al. 2013). Evolutionary studies of the *M. oleifera* genome revealed a recent surge of plastid to nucleus gene duplications that led to massive amounts of plastid DNA in the *Moringa* nuclear genome, representing 4.71%, the largest reported so far (Ojeda-López et al. 2020).

7.9 Conclusion

Moringa oleifera natural populations are broadly distributed throughout Asian tropical and subtropical regions, with rich genetic diversity in their morphology, nutritional and other useful traits. Geographic populations of *Moringa* species have varied levels of genetic relatedness and distinct genetic parameters. There is a need for concentrated biological, phylogenetic, genomic and biochemical studies on this species. Conservation efforts of Moringa as a genetic resource should be initiated across Asia to prevent extinction. The study of polymorphism and allelic frequencies revealed significantly observed and expected heterozygosity among the sampled populations across the globe, which indicated significant genetic diversity. The combined results of the phylogenetic tree, population structure analysis, and PCA indicated significant clustering across the populations of *Moringa oleifera* from different regions of the world. However, there is a need for further studies with larger collections to assess genetic diversity and eventually conserve the Moringa populations.

References

Abubakar BY, MuArsquo S, Khan AU, Adamu AK (2011) Morpho-anatomical variation in some accessions of *Moringa oleifera* Lam. from Northern Nigeria. Afr J Plant Sci 5(12):742–748

Acevedo-Rodríguez P, Strong MT (2012) Catalogue of seed plants of the West Indies. Smithsonian Contributions to Botany

Adu-Dapaah H, Osei-Bonsu I, Oduro I, Asiedu J (2015) Recent advances in production, processing and utilization of Moringa oleifera in Ghana. In: International symposium on Moringa, vol 1158, pp 179–186

Anjorin TS, Ikokoh P, Okolo S (2010) Mineral composition of *Moringa oleifera* leaves, pods and seeds from two regions in Abuja, Nigeria. Int J Agric Biol 12:431–434

Anwar F, Bhanger MI (2003) Analytical characterization of *Moringa oleifera* seed oil grown in temperate regions of Pakistan. J Agric Food Chem 51(22):6558–6563

Anwar F, Latif S, Ashraf M, Gilani AH (2007) *Moringa oleifera*: a food plant with multiple medicinal uses. Phytother Res 21(1):17–25

Arora DS, Onsare JG, Kaur H (2013) Bioprospecting of Moringa (Moringaceae): microbiological perspective. J Pharmacogn Phytochem 1(6):193–215

Asante WJ, Nasare IL, Tom-Dery D, Ochire-Boadu K, Kentil KB (2014) Nutrient composition of Moringa oleifera leaves from two agro ecological zones in Ghana. Afr J Plant Sci 8(1):65–71

Bhatia S, Othman Z, Ahmad AL (2007) Pretreatment of palm oil mill effluent (POME) using *Moringa oleifera* seeds as a natural coagulant. J Hazard Mater 145(1–2):120–126

Chang Y, Liu H, Liu M, Liao X, Sahu SK, Fu Y, Song B, Cheng S, Kariba R, Muthemba S, Hendre PS (2019) The draft genomes of five agriculturally important African orphan crops. Gigacience 8(3):giy152

Csurhes S, Navie S (2016) Horseradish tree: *Moringa oleifera*. Queensland Government, Brisbane, AUS.

da Silva AV, dos Santos AR, da Silva LA, Feitosa RB, Almeida CS, da Silva GM, Rangel MS (2012) Moringa genetic diversity from germplasm bank using RAPD markers. Trop Subtrop Agroecosystems 15 (1):31–39

Estrella MC, Jacinto Bias III, V, David GZ, Taup MA, (2000) A double-blind, randomized controlled trial on the use of malunggay (*Moringa oleifera*) for augmentation of the volume of breastmilk among non-nursing mothers of preterm infants. Phillipp J Pediatr 49(1):3–6

Falowo AB, Mukumbo FE, Idamokoro EM, Lorenzo JM, Afolayan AJ, Muchenje V (2018) Multi-functional application of *Moringa oleifera* Lam. in nutrition and animal food products: a review. Food Res Int 106:317–334

FAO (2020). http://www.fao.org/traditional-crops/moringa/en/

Godino M, Arias C, Izquierdo MI (2015) Moringa oleifera: potential areas of cultivation on the Iberian Peninsula. In: International symposium on Moringa, vol 1158, pp 405–412

Gopalakrishnan L, Doriya K, Kumar DS (2016) *Moringa oleifera*: a review on nutritive importance and its medicinal application. Food Sci Hum Well 5(2):49–56

Hassanein AM (2018) Nutritional, chemical and molecular characterisation of *Moringa oleifera* Lam. and *Moringa peregrina* (Forssk.) Fiori genotypes. J Hortic Sci Biotechnol 93(5), 500–5009

Hsu R, Midcap S, Arbainsyah DW (2006) *Moringa oleifera* medicinal and Economic uses. International course on Economic botany, National Herbarium, Leiden, The Netherlands

IMGC (2017) International Moringa Germplasm Collection. https://moringaceae.org/index.html

Islam MA, Sheikh A, Waterman C, Hosenuzzaman M (2020) Morphology, pod yield and nutritional quality of two cultivars of Moringa (*Moringa oleifera*) in Bangladesh. Indian J Sci Technol 13(37):3893–3903

Khalafalla MM, Abdellatef E, Dafalla HM, Nassrallah AA, Aboul-Enein KM, Lightfoot DA, El-Deeb FE, El-Shemy HA (2010) Active principle from *Moringa oleifera* Lam leaves effective against two leukemias and a hepatocarcinoma. Afr J Biotechnol 9(49):8467–8471

Kumar Ganesan S, Singh R, Choudhury DR, Bharadwaj J, Gupta V, Singode A (2014) Genetic diversity and population structure study of drumstick (Moringa oleifera Lam.) using morphological and SSR markers. Ind Crop Prod 60:316–325

Leone A, Spada A, Battezzati A, Schiraldi A, Aristil J, Bertoli S (2015) Cultivation, genetic, ethnopharmacology, phytochemistry and pharmacology of *Moringa oleifera* leaves: an overview. Int J Mol Sci 16 (6):12791–12835

Magat SS, Raquepo MC, Pabustan CD (2009) Mineral macro-nutrients, micro-nutrients and other elements in leaves of malunggay plant (*Moringa oleifera*) sampled in some locations in The Philippines. Technology-Advisory Notes

Mallenakuppe R, Homabalegowda H, Gouri MD, Basavaraju PS, Chandrashekharaiah UB (2015) History, taxonomy and propagation of *Moringa oleifera*-a review. Crops J 3(3.28):3–15

Mgendi MG, Manoko MK, Nyomora AM (2010) Genetic diversity between cultivated and non-cultivated *Moringa oleifera* Lam. provenances assessed by RAPD markers/ *Moringa oleifera*'nin ekili olan ve olmayan kökenlerinin genetik çesitliliginin RAPD markörleri ile degerlendirilmesi. Mol Cell Biol 8(2):95

Moyo B, Masika PJ, Hugo A, Muchenje V (2011) Nutritional characterization of Moringa (*Moringa oleifera* Lam.) leaves. Afr J Biotechnol 10 (60):12925–12933

Muluvi GM, Sprent JI, Soranzo N, Provan J, Odee D, Folkard G, McNicol JW, Powell W (1999) Amplified fragment length polymorphism (AFLP) analysis of genetic variation in *Moringa oleifera* Lam. Mol Ecol 8 (3):463–470

Muluvi GM, Sprent JI, Odee D, Powell W (2004) Estimates of outcrossing rates in *Moringa oleifera* using Amplified fragment length polymorphism (AFLP). Afr J Biotechnol 3(2):145–151

Nouman W, Anwar F, Gull T, Newton A, Rosa E, Domínguez-Perles R (2016) Profiling of polyphenolics, nutrients and antioxidant potential of germplasm's leaves from seven cultivars of *Moringa oleifera* Lam. Ind Crop Prod 83:166–176

Ohri D, Kumar A (1986) Nuclear DNA amounts in some tropical hardwoods. Caryologia 39(3–4):303–307

Ojeda-López J, Marczuk-Rojas JP, Polushkina OA, Purucker D, Salinas M, Carretero-Paulet L (2020) Evolutionary analysis of the *Moringa oleifera* genome reveals a recent burst of plastid to nucleus gene duplications. Sci Rep 10(1):1–5

Olson ME (2002) Combining data from DNA sequences and morphology for a phylogeny of Moringaceae (Brassicales). Syst Syst Bot 27(1):55–73

Palada MC, Ebert AW, Yang RY, Chang LC, Chang J, Wu DL (2017) Progress in research and development of Moringa at the World Vegetable Center

Paliwal R, Sharma V, Sharma S (2011) Elucidation of free radical scavenging and antioxidant activity of aqueous and hydro-ethanolic extracts of *Moringa oleifera* pods. Res J Pharm Tech 4(4):566–571

Pandey A, Pradheep K, Gupta R, Nayar ER, Bhandari DC (2011) 'Drumstick tree'(*Moringa oleifera* Lam.): a multipurpose potential species in India. Genet Resour Crop Evol 58(3):453–460

Patricio HG, Palada MC (2015) Adaptability and horticultural characterization of different Moringa accessions in Central Philippines. In: International symposium on Moringa, vol 1158, pp 45–54

Pirrò S, Minutolo A, Galgani A, Potestà M, Colizzi V, Montesano C (2019) Bioinformatics prediction and experimental validation of MicroRNAs involved in cross-kingdom interaction. J Comput Biol 23 (12):976–989

Popoola JO, Bello OA, Obembe OO (2016) Phenotypic intraspecific variability among some accessions of drumstick (*Moringa oleifera* Lam.). Can J Appl Sci 10 (1):3681–3693

Popoola JO, Bello OA, Olugbuyiro J, Obembe OO (2017) Simple sequence repeats (SSR) analysis of genetic intraspecific relationships of *Moringa oleifera* populations from Nigeria. Sci Int 29(3):645–657

Popoola J, Igwe D, Jegede O, Iwu V, Adegbite A, Omonhinmin C (2019) Agronomic practices, genetic diversity and population structure of *Moringa oleifera* (Lam.) in Nigeria, Preprints.org

PROTA (2017) PROTA4U web database. Wageningen, Netherlands: Plant Resources of Tropical Africa. https://www.prota4u.org/database/

Raja S, Bagle BG, More TA (2013) Drumstick (*Moringa oleifera* Lam.) improvement for semiarid and arid ecosystem: analysis of environmental stability for yield. J Plant Breed Crop Sci 5(8):164–170

Rajalakshmi R, Rajalakshmi S, Parida A (2019) Genetic diversity, population structure and correlation study in *Moringa oleifera* Lam. using ISSR and SRAP markers. Proc Natl Acad Sci India Sect B Biol Sci 89 (4):1361–1371

Ramachandran C, Peter KV, Gopalakrishnan PK (1980) Drumstick (*Moringa oleifera*): a multipurpose Indian vegetable. Econ Bot 1:276–283

Rufai S, Hanafi MM, Rafii MY, Ahmad S, Arolu IW, Ferdous J (2013) Genetic dissection of new genotypes of drumstick tree (*Moringa oleifera* Lam.) using random amplified polymorphic DNA marker. BioMed Res Int

Saini RK, Saad KR, Ravishankar GA, Giridhar P, Shetty NP (2013) Genetic diversity of commercially grown *Moringa oleifera* Lam. cultivars from India by RAPD, ISSR and cytochrome P 450-based markers. Plant Syst Evol 299(7):1205–1213

Sánchez NR, Ledin S, Ledin I (2006) Biomass production and chemical composition of *Moringa oleifera* under different management regimes in Nicaragua. Agrofor Syst 66(3):231–242

Sánchez-Machado DI, Núñez-Gastélum JA, Reyes-Moreno C, Ramírez-Wong B, López-Cervantes J (2010) Nutritional quality of edible parts of *Moringa oleifera*. Food Anal Methods 3(3):175–180

Shahzad U, Khan MA, Jaskani MJ, Khan IA, Korban SS (2013) Genetic diversity and population structure of *Moringa oleifera*. Conserv Genet 14(6):1161–1172

Siddhuraju P, Becker K (2003) Antioxidant properties of various solvent extracts of total phenolic constituents from three different agroclimatic origins of drumstick tree (*Moringa oleifera* Lam.) leaves. J Agric Food Chem 51(8):2144–2155

Stephenson KK, Fahey JW (2004) Development of tissue culture methods for the rescue and propagation of endangered *Moringa spp.* germplasm. Econ Bot 58(1): S116–124

Suarez M, Entenza JM, Doerries C, Meyer E, Bourquin L, Sutherland J, Marison I, Moreillon P, Mermod N (2003) Expression of a plant-derived peptide harboring water-cleaning and antimicrobial activities. Biotechnol Bioeng 81(1):13–20

Tak S, Maurya IB (2015) Genetic diversity of Moringa oleifera Lam. in Rajasthan, India. In: International symposium on Moringa, vol 1158, pp 71–78

Vogt K (1996) A field worker's guide to the identification, propagation and uses of common trees and shrubs of dryland Sudan

Waterman C, Cheng DM, Rojas-Silva P, Poulev A, Dreifus J, Lila MA, Raskin I (2014) Stable, water extractable isothiocyanates from *Moringa oleifera* leaves attenuate inflammation in vitro. Phytochemistry 03:114–122

Wu JC, Yang J, Gu ZJ, Zhang YP (2010) Isolation and characterization of twenty polymorphic microsatellite loci for *Moringa oleifera* (Moringaceae). HortSciences 45(4):690–692

Yang R Y, Chang LC, Hsu JC, Weng BB, Palada MC, Chadha ML, Levasseur V (2006) Nutritional and functional properties of Moringa leaves–From germplasm, to plant, to food, to health. Moringa leaves: Strategies, standards and markets for a better impact on nutrition in Africa. Moringanews, CDE, CTA, GFU Paris

Tissue Culture and Genetic Engineering in Moringa

8

N. Manikanda Boopathi, J. D. Harshith, V. P. Santhanakrishnan, and M. Raveendran

Abstract

In vitro and aseptic culturing of cells, tissues, organs, and their components in artificial growth media and environments is referred as plant tissue culture, which has applications in basic and applied studies as well as in commercial production of quality planting materials. Role of plant tissue culture tools in Moringa has several folds of applications: understanding cellular activities (including cell growth and development, cytogenetics, metabolic process and nutrition, morphogenesis, embryogenesis, and pathology), manipulating plant genetic architecture and improvement, production of disease-/pathogen-free plants, in vitro germplasm storage, high-throughput clonal propagation, and product development (such as secondary metabolite production, natural color isolation, etc.). Besides, establishment of efficient Moringa in vitro culture is the critical basic requirement of genetic transformation (which provide potential of transferring gene (originated from any organism) that governs the expression of desirable traits). This chapter provides comprehensive information on Moringa tissue culture and its applications besides providing the future prospects.

8.1 Scope for Plant Tissue Culture in Moringa

Moringa is truly a tropical plant, which is largely grown in dry arid and semi-arid tracts. These areas are found to be the most suitable for its cultivation to obtain a profitable yield with minimum efforts. By realizing the huge demand for Moringa plants for commercial cultivation, all the possible methods to propagate Moringa have been tried. Conventionally, macropropagation methods such as asexual (or clonal) propagation (i.e., limb cuttings and air layering) and sexual propagation (i.e., seed multiplication) are frequently used to propagate Moringa.

Asexual or clonal propagation is usually described as production of several numbers of genetically identical copies of a cultivar by asexual means. Apomixis (which is defined as formation of seed without fertilization and meiosis) and/or vegetative propagation (reproduction of progenies from vegetative parts of the plants) are noticed as natural processes that help to induce clonal propagation.

Moringa propagation through limb cuttings is usually chosen by the farmers as it can be simply performed by them with the available limited resources. In addition, limb cutting from healthy crop develops roots very easily. Usually, cuttings

N. M. Boopathi (✉) · J. D. Harshith · V. P. Santhanakrishnan · M. Raveendran
Department of Plant Biotechnology, Center for Plant Molecular Biology & Biotechnology, Tamil Nadu Agricultural University, Coimbatore, India
e-mail: nmboopathi@tnau.ac.in

N. M. Boopathi et al. (eds.), *The Moringa Genome*, Compendium of Plant Genomes,
https://doi.org/10.1007/978-3-030-80956-0_8

of fairly large size stems (with a size of 1–1.35 m length and 14–15 cm girth) are planted in moist soil (preferably during the rainy season, June–August, in India) to easily establish roots, and it can attain a 2–3 m height within 3–4 months. Alternatively, in selected regions of Tamil Nadu, India, unproductive trees are trimmed to leave main stem with a length of ~1 m and maintained to grow 2–3 shoots. Once the shoots attain sufficient growth, shoots that are 2 m in length and 4–5 cm in diameter are used as additional planting materials (Ramachandran et al. 1980).

Similarly, shield budding is another successful vegetative propagation method as the budded trees start to yield after 5 months. In certain traditional varieties, this strategy was found to constantly provide significant yield at least for 13 years. At this point, it is also worth to mention that air layering, another imperative commercial and remunerative Moringa propagation technology is followed in Tamil Nadu, India. Air layering has been done on the live stem of Moringa by slicing or wounding the small part and wrap around a wounded section of the stem using moist sphagnum moss or well-composed farm yard manure.

However, all the above macropropagation technologies are useful for small-scale farming, and as you can observe from the aforementioned macropropagation protocols, large numbers of Moringa plants needed for high-throughput cultivation, cannot be multiplied using these methods within a given period of time and resources. Further, other limitations of these macropropagation methods are (i) trees from stem cuttings have shallow root system (and hence, frequently experiences water stress-related production constraints), (ii) reduces the total life span of the trees, and (iii) more importantly, results in poor yield (Islam et al. 2005). Though macropropagation methods are simple and successful for tree propagation, the current trend in Moringa cultivation for the production of leaf biomass forced to find other alternative methods of high-throughput propagation, such as seeds.

Therefore, besides vegetative propagation, Moringa is also propagated sexually using seeds. Usually, 50–90% of the fresh Moringa seeds germinate and it germinates 7–30 days after sowing. It should also be noted that Moringa seed lose its viability, if they are stored for more than 2 months. Sharma and Raina (1982) reported loss of Moringa seed germination at the rate of 50%, 48%, and 7.5% when they are sowed after 1, 2, and 3 months, respectively, of seed collection due to its high oil content (and also in certain cases, due to storage pest infestation). Thus, long-term storage of Moringa seeds critically affects the germination process, and hence, keeping the seeds even for the next season may have adverse effect on the germination.

In order to overcome these limitations, various priming strategies (such as hardening, hydropriming, osmopriming, matripriming, osmohardening, and hormonal priming) have been developed to rejuvenate Moringa seeds. Such priming treatments reduce the time required to germinate and thus increase the germination percentage or emergence index. Subsequently, increase in seed emergence leads to uniform crop stand, which ultimately results into increased yield. Moringa leaf extract is rich in cytokinin, and its application as a seed priming tool was also exploited in several crops, including Moringa. Nouman et al. (2012) found that hydropriming (soaking in water for 8 h) has promoted seed emergence, shoot vigor, and chlorophyll b contents, while priming with Moringa leaf extract for 8 h resulted in vigorous root development and enhanced levels of chlorophyll a and minerals in Moringa leaves.

Besides the problem of lower germination percentage, plants raised from seeds leads to non-uniform plants and produce fruits of inferior quality, and more imperatively, plants have invariably showed genotypic and phenotypic variations (Salem 2015). This is mainly due to the fact that Moringa is a cross-pollinated crop; the variation largely occurs owing to the segregation of undesirable traits in the subsequent generations.

Therefore, an affordable method (that can support high-throughput or large-scale production of quality Moringa planting materials) is required, and micropropagation or plant tissue culture is proposed as a method of choice at this junction. Vegetative propagation of plants are now generally achieved by employing tissue culture or in vitro culture strategies, which involves aseptic culturing of tissues or cells under artificial chemical and physical conditions to produce huge numbers of true-to-type plantlets with relatively short time and space. This method is simply referred as micropropagation. Indeed, in several horticultural crops (such as Orchids), micropropagation is recognized as the merely viable method of propagation that has commercial value.

8.2 In Vitro Culture

In vitro culture is a cellular life process that occurs in a test tube, culture dish, or elsewhere outside a living organism. In plant system, such process is also referred as plant tissue culture (PTC), and it offers numerous applications for Moringa propagation as well as its genetic improvement. Applications of PTC in Moringa plants includes

(i) swift production of doubled haploids that lead to develop homozygous plants from the immature wide hybridized embryos with lesser time and smaller space,
(ii) propagation of endangered *Moringa* spp., which will be used as a valuable resources for genetic and physiological studies,
(iii) evolving novel genetic variation,
(iv) inducing mutagenesis and selection of cell lines for specific traits (such as production of useful secondary metabolites),
(v) exploring protoplast fusion that promotes wide hybridization and genetic transformation (explained in detail in later part of this chapter), and
(vi) high-throughput multiplications of desired Moringa varieties or ecotypes using micropropagation strategy.

Micropropagation is an outstanding strategy for both conservation of endangered plant species and production of disease-free plants. As an effective alternative strategy to the traditional propagation methods, micropropagation supports supply of quality planting materials throughout the year with minimum cost as it employs limited space and time.

8.3 Micropropagation in Moringa

Micropropagation refers to large-scale aseptic and in vitro production of true-to-type complete plants using small part of the mother plants (referred as explant), and there are three types of micropropagation:

(1) *Somatic embryogenesis*: which refers to formation of structures comprising shoot and root that are connected by a closed vascular system (this process resembles histology, physiology, and biochemistry of zygotic embryos),
(2) *Adventitious shoot production:* This is with reference to the de novo development of shoot meristem which are arising from callus or other organized tissues such as epidermal or subepidermal cells, and
(3) *Axillary shoot production*: This process involves production of additional shoots from axillary buds and meristems under in vitro conditions.

In this chapter, micropropagation is used to refer the mass multiplication of large true-to-type plants for commercial purposes, mainly for cultivation. Though the word micropropagation is also used to describe development of shoots from transgenic callus, preservation of male sterile and other pre-breeding plant materials, development of doubled haploid plants, etc., they were not referred in this chapter. Similar to its effective and extensive applications in several horticultural crops, micropropagation has also been shown to be flourishing in Moringa. Since Moringa cultivation requires clonal propagation of thousands of propagules to cover large acreage, micropropagation in Moringa is fundamentally considered as extensive scaling-up technology.

In order to increase the yield and biotic and abiotic stress resistance, exploring the concept of heterosis and hybrids has long been explored in crop plants. Novel genes can be introduced to the elite cultivars by employing single- or double-cross hybrids using either controlled pollination or transformation technology (which is explained below in Sect. 8.8). Thus, by sequentially crossing four lines, as shown in Fig. 8.1, a double-cross system can be established, and the consequential hybrid can be micropropagated.

8.4 Basics and Practice of Moringa Micropropagation

Micropropagation is proposed as a viable commercial method to produce large numbers of Moringa propagules, since there are insufficient seeds or cuttings available for planting. The earliest report on Moringa micropropagation was presented by Mohan et al. (1995), and they successfully multiplied *M. oleifera* from immature embryo-, seedling-, and mature tree-derived explants. Though successful reports on micropropagation have long been reported elsewhere (Islam et al. 2005; Stephenson and Fahey 2004; Xiang et al. 2007; Marfori, 2010; Saini et al. 2012), till now it has not been established as a complete commercial venture. This section provides basic procedure of micropropagation in Moringa (Fig. 8.2) besides explaining its other potential applied aspects.

The fundamental requirement for Moringa micropropagation is selection of appropriate explants. Explants (defined as any part of the Moringa plant) have the ability to grow into complete plantlets under aseptic conditions in an artificial nutrient medium. This property of the explant is referred as *totipotency* which forms the basis of micropropagation.

Published reports have frequently used nodal explants obtained from non-aseptic sources such as young seedlings or mature plants. In general, a nodal segment of 2–4 cm is taken from the mother Moringa plant and sterilized for a short period with sterilizing agents such as 0.1% mercuric chloride (w/v) for 2 min and 20% sodium hypochlorite (v/v) for 10 min. Residual sterilizing agents are removed by rinsing the explants with sterile distilled water for three times. It is preferable to perform this surface sterilization of explants inside the laminar flow hood.

Alternatively, aseptic explants can also be excised from the in vitro grown Moringa seedlings as briefly described hereunder: using the above sterilizing agents, uniformly matured Moringa seeds can be surface-sterilized. This is followed by aseptic removal of seed coats and further sterilization with 20% sodium hypochlorite (v/v) (for 5 min), and finally, the residual sodium hypochlorite are removed by rinsing in sterile distilled water at least for three times. Such sterilized seeds are inoculated aseptically in Murashige and Skoog basal medium (MS salts; Murashige and Skoog 1952) containing 30 g L^{-1} sucrose (to provide instant energy) and 5 g L^{-1} agar (which provides semi-solid environment for the growth of plants). The pH of the medium is adjusted to 5.8, and the medium is distributed to culture bottles (@ 40 ml each) which are previously sterilized by autoclaving at 121 °C for 20 min. The inoculated nodal explants are kept in the dark at 27 ± 1 °C for 15 to 30 days

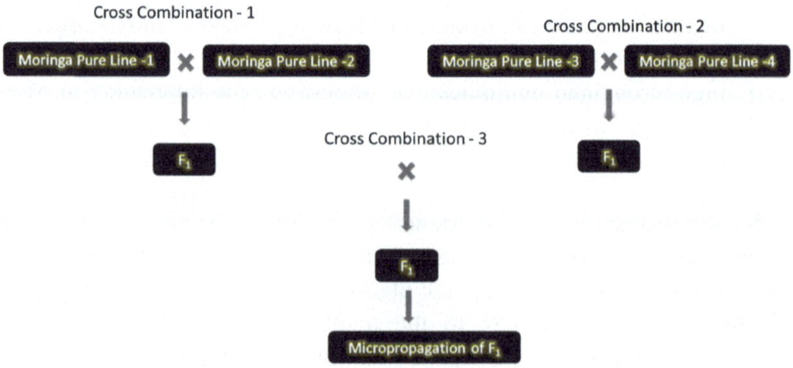

Fig. 8.1 A micropropagation scheme for large-scale production of double-cross Moringa line

Fig. 8.2 Schematic workflow for micropropagation of Moringa cultivar PKM1

(depends on the Moringa variety/ecotype, cultural conditions, etc.), until they germinate. Once the seedlings got germinated, they are transferred under continuous light provided by cool white-fluorescent tubes (at 2,000-lx intensity). Germinated seedlings consisting of three to four nodes (which can be obtained within 3–4 weeks after inoculation) will be used in the downstream process as detailed below (Fig. 8.2).

8.4.1 Multiple-Shoot Induction

Nodal explants that were prepared either from field grown plants or grown under aseptic conditions, as above, will be subjected to induce multiple shoots by employing the following protocol: Nodal explants are planted on a multiple shoot induction medium (comprising MS salts supplemented with benzyl adenine (BA) @ 4.44 μM) for multiple axillary shoot formation.

Plant growth regulator present in the MS medium alone is not sufficient for effective development of shoots and roots, and hence, MS is supplemented with cytokinin and auxins. Usually, addition of cytokinin in the MS medium, such as BA, induces shoot formation, and auxins, such as IAA and IBA, enhance root formation in plant tissue culture protocol.

Micro shoots obtained are repeatedly subcultured (minimum of four and maximum of eight times) in MS basal medium accompanied with 4.44 μM BA at 20–30 days interval when it attained sufficient growth (usually 8–12 cm height).

8.4.2 Rooting of Micropropagated Moringa Shoots

Once sufficient number and growth of axillary shoots were developed from nodal sections, they need to be placed in a root induction medium (which is a MS medium supplemented with 2.85 μM indole-3-acetic acid (IAA) and 4.92 μM indole-3-butyric acid (IBA)). This change in media composition enhances in vitro rooting of micropropagated Moringa shoots within 7–10 days. For some of the Moringa ecotype, such high salt content of the medium may restrict growth and development of root. In that case, use of half-strength MS medium supplemented with 0.5 mg L^{-1} IAA will be useful to induce roots. Once roots are developed (in general, 30–45 days are sufficient to develop prominent roots), such rooted plantlets are transferred to the soil for hardening.

Gupta et al. (2020) reported that complete removal of the surface contaminants was possible with the treatment of decoated seeds with 1% (w/v) Bavistin for 50 min, 0.33% (w/v) streptocycline for 30 min, and 0.1% (w/v) $HgCl_2$ for 3.5 min and maximum seed germination (89.13%) was obtained on quarter-strength MS medium. Complementary effect of plant growth regulators on axillary bud proliferation in Moringa has also been assessed by them, and it was noticed that culturing of nodal segments on MS medium combined with BA at 3 mg L^{-1} proliferated multiple shoots (maximum of 18 shoots per explants was reported in this study). Strikingly, it has been shown that lesser amounts of BA (0.5 mg L^{-1}) along with IAA (0.5 to 2 mg L^{-1}) or Kn (0.5 to 5 mg L^{-1}), showed combined effort in the shoot morphogenesis. In addition, the maximum of 100% rooting competence was accomplished with half-strength MS medium comprising either IAA or IBA. For example, average root length of 7.3 ± 0.8 cm can be induced using half-strength MS supplemented with IBA (0.1 mg L^{-1}).

8.4.3 Primary and Secondary Hardening

Acclimatizing the aseptically grown Moringa plants to the farmer's field condition is called as hardening. Initially, the rooted tissue culture plants are planted in plastic bags containing autoclaved mixture of soil, sand, and vermicompost (@ 3:1:1 v/v), and this process is called as primary hardening. During the primary hardening process, the plants require adequate care: i) plants should be kept under limited sunlight under greenhouse at surrounding temperature of ~25–28 °C and covered with transparent and perforated polythene bags, and ii) watering has to be done regularly, and if possible, it may be supplied as mist spray.

The plants can be kept under the above setup for 15 days and after that the polythene covers can be removed. However, the plants have to be maintained under the greenhouse for additional 15–20 days, and this process is referred as secondary hardening. Finally, secondary hardened plantlets are transferred to the farmer's field, and their performance is evaluated in comparison with the regularly propagated Moringa plants.

If there is any deviation in the performance of the micropropagated Moringa, it is advisable to reduce the number of subculturing or change the explant source. It was noticed that micropropagated plants were found to be nutritionally richer when compared with conventionally grown plants and Moringa also has witnessed such trends (Saini et al. 2012). Greater amount of phytochemicals and nutrients in the tissue culture-derived plants might be due to progressive shoot development of the tissue cultured plant (i.e., vigorous vegetative growth) that prevails during the planting which helps to reach better growth and development.

8.4.4 Clonal Fidelity Analysis and Confirmation of Authenticity of Micropropagated Moringa

Plantlets

The authenticity of micropropagated Moringa propagules can be typically verified by using molecular markers. Somaclonal variation is the rare phenomenon found during the micropropagation and other in vitro culture procedures. It is generally referred to changes in genetic elements (i.e., DNA) of plant tissue due to the micropropagation and such changes are attributed to:

(1) Employing several sources of explants,
(2) Physical and Chemical status of cultural conditions,
(3) Regeneration methods employed during in vitro culturing, and
(4) Long course of subcultural practices.

Additionally, induction of several chromosomal abnormalities owing to excess concentration of plant growth hormones is also found to curtail production of true-to-type plants. Thus, when Moringa micropropagation is employed to produce true-to-type quality plants, somaclonal variation is recognized as a major concern. Therefore, evaluating the genetic stability of micropropagated Moringa becomes mandatory.

In order to test the genetic authenticity of the micropropagated plants, simple and robust polymerase chain reaction (PCR)-based molecular markers such as Inter Simple Sequence Repeats (ISSR), Start Codon Targeted (SCOT) marker, Randomly Amplified Polymorphic DNA (RAPD), and Amplified Fragment Length Polymorphism (AFLP) can be employed (Boopathi 2020). If the banding patterns (that are produced by a set of random primers used in the above marker classes) are found to be identical, then it is used to confirm the clonality of all the investigated micropropagated plants.

It is generally recommended that use of more than one type of molecular marker could increase the efficiency of the genetic fidelity assessment (Behera et al. 2019) of micropropagated Moringa plants.

The procedure for testing the genetic fidelity using RAPD marker (Boopathi 2020) is briefly provided here: Extract the DNA from the leaf tissues of micropropagated plantlets that are randomly selected from each batch of in vitro culture using cetyl trimethylammonium bromide (CTAB) method (Doyle 1991) or using commercial kits. Select a set of 20–25 random primers (the sequence of each primer can be obtained from the website of Operon Technologies, Alameda, CA) and artificially synthesize it. PCR can be performed by preparing a total cocktail volume of 25 μl (by mixing 1 μl of the DNA sample (which is prepared as 50 ng/μl concentration), 2.5 μl of 10X *Taq* Buffer, 200 μM of each dNTPs, 2.5 mM MgCl2, 0.75 U of *Taq* polymerase, 0.8 μM primer, and make up to 25 μl with deionized water) using the commercially available Thermal Cyclers with the following amplification conditions: initial denaturation at 95 °C for 5 minutes to denature the entire Moringa DNA. This is followed by programming 35 cycles of 95 °C for 50 s, 37 °C for 45 s, and 72 °C for 45 s. Finally, a final extension of PCR products at 72 °C for 5 minutes is also programmed to complete the amplification of all the PCR products.

The PCR products are analyzed using 1.5% (w/v) agarose gel electrophoresis. The banding pattern is documented using a Gel Documentation System after staining the gel with ethidium bromide. Verification of all the micropropagated propagules that have produced the same banding pattern as that of mother plant confirms the fidelity of the micropropagated products. Variations in the banding pattern denote there is a somaclonal variation and the investigated micropropagation lot needs to be discarded to ensure that every micropropagated plants are true-to-type to the mother plant.

Fidelity of the uniform genetic content of micropropagated plants of drumstick tree was demonstrated with different molecular markers in several occasions. For example, Avila-Treviño et al. (2017) employed Randomly Amplified Microsatellite Polymorphism (RAMP) markers. They have also demonstrated that somaclonal variation was minimal in shoot tip-based clonal propagation of *M. oleifera* by validating true-to-type nature of in vitro raised plantlets using RAMP.

Similarly, clonality of the Moringa was assessed by Gupta et al. (2020) using RAPD and ISSR markers and found that there was an amplification of distinct, identical, and reproducible bands. Upon scrutinizing the micropropagated plants with RAPD and ISSR markers, they have found that there was a high monomorphism (95.39%) among the investigated plants. This confirms the high genetic stability and clonal fidelity among the tissue cultured plants.

8.5 Chief Applications of Micropropagation in Moringa

1. Mass propagation of genetically uniform planting materials: Instead of obtaining

~ 10,000 practically utilizable plants per year through vegetative propagation (such as limb cuttings that measures ~ 1 m with 12–15 cm girth) from a hectare of Moringa field, micropropagation can provide more than 1 million genetically uniform plants per year from 1000 s of explants (preferably nodal segments that measures 2–3 cm).
2. In vitro culture can be initialized from small parts of plants with relatively smaller space: 100 m^2 space of micropropagation unit (including culture room and greenhouse) is sufficient for annual production of 1 million plants.
3. Mass multiplication of disease- and virus-free plantlets make things easier during the international exchange of plants, besides offering huge reduction in cultivation cost involved in management of seed- and/or soil-borne pest and diseases during the establishment of Moringa garden.
4. Micropropagation also facilitates introduction of disease-free elite cultivars to the new production zone, elsewhere in the world.
5. Sterile hybrids or multiple cross hybrids (Fig. 8.1) can be easily multiplied using vegetative tissues using micropropagation.
6. Conservation of endangered *Moringa* spp., and their ecotypes is need of the hour. Cryopreservation of Moringa in vitro cultures will be very much useful to store such germplasm for long periods.
7. Genetic improvement of Moringa can be completed with lesser time than it requires usually through conventional breeding methods as micropropagation shortens the breeding cycle.

8.6 Key Issues in Micropropagation in Moringa

Though commercial viability of micropropagation in Moringa has been established, below points encapsulates key issues of micropropagation in Moringa:

1. Success in Moringa micropropagation is largely depending on type and source of explants and type of cytokinin and auxin. There is no simple universal combination of these factors for Moringa micropropagation, and they need to be standardized across the laboratories and ecotypes (Table 8.1). This requires significant initial investment in infrastructure development due to expensive laboratory equipment and service.
2. It has often been found that Moringa explants are continuously exposed to stress under in vitro culture conditions and that led to induce certain antioxidant enzymes (such as superoxide dismutase, catalase, and peroxidase) in the explants to combat such stresses. This is an extra burden to the Moringa growth and development under in vitro conditions and all the ecotypes are not responding to this stress uniformly.
3. Explant variations in the efficiency of somatic embryogenesis: Owing to the variations in endogenous plant hormones among the explants that are derived from different sources, all such explants are not uniformly responding to the tissue culture conditions. Therefore, it is recommended to use explants derived from in vitro grown Moringa (see Sect. 8.4) as they have superior ability to get into organogenesis when compared with explants excised from farmer's field (i.e., in vivo grown plants).
4. Degree of somaclonal variation: As explants derived from different sources have different rates of in vitro regeneration potentials, genetic variation among regenerated plants is widespread in the micropropagation process. This necessitates vigorous testing of genetic fidelity of the in vitro plants, and it is, thus, highly essential to obtain true-to-type plants. Such endeavor further enhances the efforts, time, and money that involved in micropropagation of Moringa.
5. Cost of production: In order to find the finest in vitro culture conditions for Moringa micropropagation, it requires several costlier studies and efforts (such as ascertaining the

Table 8.1 Standardization of culture conditions with respect to different explants of Moringa (only selected references are shown here to represent the need for standardization of micropropagation in Moringa)

Explants employed	Sterilizing agent employed	Media and supplements	Remarks or additional works employed	References
Stem cuttings from the in vitro germinated seeds	70% ethanol for 2 min + 15 min in 25% Clorox™ bleach	MS + 1 mg L^{-1} BAP, 1 mg L^{-1} GA3, 0.25% activated charcoal for shoot multiplication Half-strength MS + 0.5 mg/L NAA (93.1 mM) for primary and secondary root development	The success rate was 73% and the multiplication rate averaged 4.7 shoots per culture	Stephenson and Fahey (2004)
Cotyledonary node of decapitated seedlings	1% sodium hypochlorite solution mixed with a drop of Tween-20 for 20 min	Full strength MS for multiple shoot induction Half-strength MS for rooting	Attempted micropropagation in *Moringa Oleifera* Lamk., *Moringa stenopetala* (Baker f.) Cufod, and *Moringa peregrine* Forssk. ex Fiori and established method to secure biodiversity; establishment of vegetative clones and avoided laborious and time-consuming usage of plant growth regulators	Steinitz et al. (2009)
Nodal segments from in vitro germinated seeds	Seeds surface-sterilized by wiping them with 95% ethanol and flaming for 30 s	MS + 2.5 μM BAP produced 4.5 axillary shoots per explant MS + 0.25 μM NAA for rooting	Only 80% plants survived	Marfori (2010)
Nodal segments from in vitro germinated seeds	Sterilizing seeds: 0.1% mercuric chloride (w/v) for 2 min and 20% sodium hypochlorite (v/v) for 10 min	MS + 4.44 μM BA produced 9.0 ± 1.0 axillary shoots per explant MS + 2.85 μM IAA + 4.92 μM IBA for 100% rooting	Eighty percent of the tissue cultured plants survived after transplanting and they were found nutritionally superior over conventionally grown plants	Saini et al. (2012)
Nodal segments obtained from pot grown seedlings	70% ethyl alcohol (v/v) for 30 s and 20% sodium hypochlorite (v/v) for 3 min	Shoot multiplication: MS + 2.5 μM BAP under mild ventilation Root Induction: MS + 4.92 μM IBA for adventitious	Under salinity, in vitro shoots showed symptoms of vitrification	Salem (2015)
Nodal segments and shoot cuttings (shoots and shoot tips)	Consequent treatment with 70% ethyl alcohol and 0.1% HgCl2 for 3 min in each step	MS + 0.55 mg L^{-1} of BAP produced 9.57 ± 2.08 shoots. Half-strength MS + 0.5 mg L^{-1} IAA for successful root formation	Impact of reactive oxygen species on vitrification was studied	Hassanein et al. (2019)

(continued)

Table 8.1 (continued)

Explants employed	Sterilizing agent employed	Media and supplements	Remarks or additional works employed	References
Nodal segments from in vitro germinated seeds	1% (w/v) Bavistin for 50 min followed by 0.33% (w/v) streptocycline for 30 min and 0.1% (w/v) $HgCl_2$ for 3.5 min	Multiple shoot proliferation: MS + 5-benzyladenine (BA) (3 mg L^{-1}) Rooting Media: half-strength MS + IAA (0.1 mg L^{-1}) + IBA (0.1 mg L^{-1})	Multiplication ratio of 18 shoots per explant and 100% rooting efficiency was reported. Genetic uniformity of the micropropagated plants was confirmed using molecular markers	Gupta et al. (2020)

type and source of explants, best concentration and type of growth regulators, and affordable and optimal media type that stimulate better organogenesis).

6. Another common problem in Moringa micropropagation is vitrification of the regenerated shoots. Vitrification (also well recognized as Hyperhydricity) is a physiological malformation found in the in vitro regenerated Moringa plants due to disproportionate hydration, little lignifications, weakened stomatal activity, and decreased mechanical strength of stems. Therefore, the regenerated Moringa plants that experiences hyperhydricity often possess poor acclimatization to the farmer's field, which is already characterized as a resource limited environment. Therefore, in order to realize the potential of the micropropagation in Moringa, problems of both vitrification and somaclonal variation have to be circumvented. To this end, Hassanein et al. (2018) proposed the addition of silver nitrate and salicylic acid in the culture medium to avoid vitrification and for durable maintenance of vigorous shoots.

8.7 Untapped Applications of Plant Tissue Culture Strategies in Moringa

Besides micropropagation, other strategies of plant tissue culture (including protoplast fusion and cybridization, exploiting desirable somaclonal variation, in vitro germplasm conservation, production of double haploids, cell suspension culture and secondary metabolite production, and fabrication of synthetic seeds) may offer numerous applications in Moringa. Though such applications have not been fully realized in Moringa, their potentials are briefly described hereunder anticipating their utility in Moringa.

8.7.1 Protoplast Fusion and Cybridization

Sexual breeding is the key process by which genetic materials are transferred from one species to another eukaryotic species. Though transfer of genetic materials has been realized among distantly related plant species, barriers such as sexual incompatibility restrict the development of complete and viable hybrids between desired and distantly related plant species. It is also well documented in Moringa that even within species and intra-varietal cross combinations, success has been found to be difficult due to several factors (including small flowers that occur in fragile tree branches). To this end, somatic hybridization resulting from fusion of unrelated cells (for example, *M. oleifera* and *M. drouhardii* or *M. ovalifolia* and *M. stenopetala*) can be employed as a novel approach to execute distant hybridization in Moringa, and fusion can be accelerated through the plasma membrane.

However, plasma membrane in plants is surrounded by strong cellulosic walls, which are further strengthened by pectin-rich matrix that tightly joins neighboring cells. Therefore, advances in animal somatic cell genetics is stronger (as there is no such boundaries in animal cells that surrounds plasma membrane) than plants. To break this barrier, in 1950, E. C. Cocking has first reported the successful removal of the plant cell walls using enzymes and demonstrated a procedure that acquired hefty numbers of protoplasts (which is actually viable naked cells).

Besides being able to fuse with each other, higher plant protoplasts can also take up foreign DNA, through their naked plasma membrane under specific chemical and physical treatments. Protoplasts also provide an experimental system for a wide range of biochemical and molecular studies ranging from investigations into the growth properties of individual cells to membrane transport.

However, utilization of these isolated protoplasts has been realized only when they were fused with each other, though they are isolated from different species. The process of fusing two differently sourced protoplasts was named as *somatic hybridization* as this technique totally bypassed the sexual means of hybridization. Another uniqueness of somatic hybridization is it also brings together cytoplasmic organelles from both the parents as opposed to the sexual reproduction (where only female parent contributes organelle genomes).

However, it should also be noted that though recombination in chloroplast genome of somatic hybrids was found to be rare as compared to mitochondrial genome (in which recombination happens frequently), but chloroplast segregation from either one of the parent would lead to unique and desirable nuclear-cytoplasmic genetic amalgamation and such process is called as *cybridization*. Thus, the cybridization will lead to the development of cybrid, which is referred as a cell line that has the nucleus of one parent and extra-nuclear genome(s) of the other parent.

Therefore, harnessing both the potentials of the cytoplasmic genes of one species and nuclear and cytoplasmic genes of another species is the key objective of the cybrid development. However, proofs on distribution of chloroplasts and mitotic segregation of plasma genes indicated that only a small proportion of the regenerated plantlets can be denoted as cybrid, and majority of the plantlets possess plasma genes of either one of the species.

Even though poor efficiency of cybrid development in plants has been reported, cybrid offers the following applications in plants including Moringa:

1. Simple and efficient transmission of plasma genes from one species into another (irrespective of the species, whether they are sexually compatible or not) within one generation.
2. Efficient revival of recombination occurred between the DNAs of parental mitochondria or chloroplasts.
3. Evolving different arrays/permutation of parental and recombinant chloroplasts with the various parental or recombinant mitochondria.
4. It is also possible to evolve a cybrid with mitochondria from one parent and chloroplast from another.
5. Cytoplasmic male sterility (CMS) has been successfully transferred from *Nicotiana tabaccum* to *N. sylvestris* and *Petuniahybrida* to *P. axillaries* using cybridization, since this process offered several advantages: (i) simple procedure as it is a single step procedure, (ii) the property of the elite cultivar is conserved as the nuclear genotype is unaltered and (iii) more importantly, all the progenies of cybrid definitely will be a CMS line.

8.7.2 Somaclonal Variation

Morphological and genetic variation found in the in vitro regenerated plants are described as somaclonal variation, and it has been found that such variations are mainly due to chromosomal rearrangement, modifications in ploidy level, simple and/or complex mutations in DNA or due to epigenetic factors such as exposure to plant growth regulators (PGRs), and prolonged culture time. Thus, there could be two kinds of somaclonal variations:

(1) Heritable or Genetic variability which is due to changes in DNA including mutations, chromosomal aberrations, and
(2) Non-heritable or Epigenetic variability that are found as transitory morphological changes duc to in vitro culture conditions but cannot be found in the next generation.

There are several other features that are also responsible for induction of somaclonal variation during micropropagation and they are comprehensively narrated below.

8.7.2.1 Physiological Causes of Variation

Physiological status of the explants, such as response to PGR in the growth medium and environmental conditions during the micropropagation protocol, may induce somaclonal variation. However, such factors are epigenetic, and hence, those variations would not be identified in subsequently subcultured plants and/or may not be found as Mendelian inheritance. For example, variation in regenerated plants were often found when the explants were grown in strong auxins such as phenoxyacetic acid (e.g., 2, 4-D, or 2, 4, 5-T).

8.7.2.2 Genetic Causes of Variation

Changes in nucleotide sequence or even at chromosomal level some time result in somaclonal variation. Though the regenerated plantlets are morphologically similar, the tissues of such plants originate from different sources of cell types. Hence, those regenerated plants are cytologically different as they regenerated from the dissimilar explants.

8.7.2.3 Biochemical Causes of Variation

Simple changes in biochemical pathway will lead to large implications in cellular process. For example, changes in carbon metabolism will result in poor photosynthesis (which may cause albinos in regenerated plants), and this will also have great impact in starch biosynthesis and carotenoid pathway. Variations due to changes in biochemical pathways are the most frequently found somaclonal variations during micropropagation.

8.7.2.4 Benefits of Somaclonal Variation

Though it is a major menace to micropropagation (see Sect. 8.5), such variations are sometimes beneficial. The importance of somaclonal variation has been recognized in the following areas:

1. Somaclonal variation can be used as a novel strategy to evolve desirable genetic variation in crop plants. For example, herbicide-resistant crop cultivars can be evolved by culturing its explants in the media mixed with the target herbicide. Other potential areas are developing plants resistant to disease or patho-toxins and tolerant to environmental or chemical stresses. On the other hand, identification of desirable somaclonal variation warrants careful experimental design, drawing systematic information, and utilizing the pre-breeding materials in the regular breeding program.
2. Understanding the mechanisms behind the somaclonal variation may open up new avenues in plant breeding strategies that aim to develop novel genetic variation in crop plants. It also provides fundamental molecular mechanism behind the somaclonal variation and why the clonal uniformity fails, even though it is believed that micropropagation produce true-to-type plants.
3. Somaclonal variation can also be explored to produce novel culture that has ability to produce large quantity of secondary metabolites (see Sect. 8.7.5).

When the objective of the tissue culture experiment is producing true-to-type plants or genetic transformation, it is important to reduce or completely minimize the somaclonal variation. To this end, the following strategy can be followed:

1. Frequent molecular assay (such as RAPD, SSR, ISSR; see Sect. 8.4.4) with the batch-wise subcultured plants and confirming the clonal fidelity when the objective is micropropagation of elite genotypes.

2. Somaclonal variation increases as the numbers of subculture, and hence, minimum numbers of subcultures are essential, and it is desirable to start fresh cultures using new explants. Besides, use of strong PGR such as 2,4-D in the culture medium may be avoided as it may induce variation frequently.

At this point, it should be noted that though somaclonal variation has great potential in Moringa genetic improvement program, it has not yet been reported.

8.7.3 Synthetic Seeds

In majority of crops, it has been found that the seeds are simple and affordable way of national and international exchange as storage and transportation of micropropagated plants require sophisticated transportation facility. On the other hand, seed propagation has several other inherent limitations (see Sect. 8.1). These limitations can be avoided when synthetic seeds are used.

Synthetic seeds are prepared by encapsulating somatic embryos with a suitable matrix (such as sodium alginate) enriched with mycorrhizae, insecticides, fungicides, and herbicides.

In India, Bhaba Atomic Research Center (BARC), Bombay, standardized and demonstrated the preparation of synthetic seeds in sandalwood and mulberry as early as in 1990. However, it has not yet been practiced in Moringa, and exploring such possibilities may have many applications as described below:

(i) Devoid of losing the viability, synthetic seeds can be kept for one year,
(ii) Synthetic seeds can be easily produced and simply exchanged among national and global organizations, and
(iii) As that of normal seeds, synthetic seeds can be grown in soil as direct sowing, and it does not require any hardening procedure as that of micropropagated planting materials.

However, production of synthetic seeds involves costlier efforts. As novel strategies and cost-reduction procedures are being developed, the cost of synthetic seed production will be reduced in the near future, and hence, it is expected that synthetic seeds will catch up Moringa commercial sector shortly.

8.7.4 In vitro *Germplasm Conservation*

With respect to seed storage for long period of time, seeds are grouped as orthodox seeds and recalcitrant seeds. The seeds that can tolerate the dehydration up to 5% (or sometimes even less than 5%) and retain the viability are referred as orthodox seeds and they can be stored in very low temperature with that moisture content.

On the other hand, seeds that possess highest moisture and cannot tolerate desiccation are grouped as recalcitrant seeds; the plant species living in tropical or subtropical regions produce such kind of seeds and such seeds can be maintained only in wet medium. Such maintenance will evade dehydration injury, and it also imperative to maintain them under reasonably warm environment as these species experience damage due to cold environment. Therefore, recalcitrant seeds (e.g., Moringa, oil palm, coconut, cacao coffee) can be stored for a short period (maximum of few months) even under appropriate moisture conditions and environment.

Another issue with recalcitrant seeds (e.g., gymnosperms and angiosperms) is their life span is lengthy and needs several years to set the seeds. Similarly, preserving the vegetatively propagated plants such as cassava, potato, and yams also have certain issues in long-term storage. Additionally, as there is continuous reduction in wild areas due to urbanization, in situ conservation is also impracticable, and efforts through ex situ conservation are also found to be complex as collection and conservation of all the ecotypes that represent the complete genetic diversity existing in the given species is a herculean task.

To circumvent the above issues, conservation through in vitro strategies has been proposed and several procedures for storing thousands of tropical plant species have been standardized. In vitro tissue culture techniques have been applied for not only to conserve recalcitrant and vegetatively propagated species for long time but also possess following other applications:

(1) High-throughput production of quality planting materials.
(2) As it involves aseptic means of production system, the plants produced from in vitro strategies are devoid of fungi, bacteria, viruses, and insect pests.
(3) It requires little space when compared with other conservation strategies.
(4) There is nil genetic erosion as the plants are maintained under controlled conditions.
(5) It drastically reduces the production cost as it involves less workforce.
(6) Conservation through in vitro tissue culture system simplifies the international exchange of germplasm as the sample size can be reduced to minimum (which facilitates easy shipment under sterile conditions).

Though the use of in vitro germplasm conservation in Moringa species has been reported earlier, its potential has not yet been fully explored in conservation of all the Moringa species and their ecotypes due to the following limitations of this strategy:

(i) Initial investments in establishing the in vitro tissue culture protocol for long-term storage and imparting skills to manpower to maintain the stock,
(ii) Cautious management of subcultured materials of Moringa as it involves skillful exercise, and
(iii) As the duration of in vitro storage increases, the threat to genetic variation among the stored Moringa subculture also increases, and hence, it greatly affects the uniformity of the genetic materials.

However, as new tools and technologies develop in plant tissue culture, such limitations will be avoided, and Moringa germplasm can be conserved for longer period for the people of coming century.

8.7.5 Doubled Haploids

It is always preferable to have homozygous lines of the cross-pollinating species and hybrids as it simplifies the procedure of selection and further multiplication of uniform plants. Moringa is a cross-pollinated crop. Development of homozygous lines in Moringa is a time-consuming and laborious process and requires at least seven to eight generations of self-pollinated progenies. Further, such effort is impossible in self-incompatible, male sterile, and tree species.

Alternatively, by employing diploidization of the haploid tissues, homozygous lines can easily be produced within a single generation. This process of producing homozygous dihaploids (or doubled haploids; DH) through plant tissue culture procedure is considered to be stable and equivalent to Fα generation of the conventional breeding strategy, and hence, it is considered as fastest method of producing homozygous lines.

In order to diploidize the pollen obtained from the target crop, colchicine is usually employed, and it is applied for this purpose in several ways: (i) immersing the pollen derived plants (also called as anther cultured plants) in colchicine, (ii) smeared as lanolin paste, (iii) introducing colchicine into the secondary buds, and (iv) feeding colchicine through roots.

It is established that colchicine induces chromosome duplication and changes in genic and non-genic elements and instabilities in chromosomes. Chromosome duplication is the widespread impact of colchicine, and hence, it has been explored in duplication of chromosomes in haploid chromosomes of regenerated plant as well as callus cells to generate homozygous and fertile diploid plants.

During this process, callus is induced by culturing the explants (pieces of vegetative parts such as stem, root, or petiole segments) in an appropriate medium. Though the very few diploid callus are found during the initial period, the frequency of obtaining desirable diploid cells will be increased as the subculturing number increases. The selected diploid calli are subjected to the regeneration process and the plants derived from this method are generally diploid. Upon confirming the diploid status, the plants can be used in the further downstream experiments.

It takes nearly a decade to release a novel variety (require 4–5 years for backcrossing and 4–5 years of inbreeding) in a routine conventional breeding program especially under hybridization strategy. However, fixing of desirable F_1 hybrids is possible by employing anther culture of the hybrid gametes within a single generation. Further, novel recombinations may also be generated during anther culture and such recombinations can also be concurrently fixed.

Despite such potential benefits, DH lines have not yet been explored in Moringa, and it is strongly believed that upcoming Moringa breeding program will explore this potential avenue.

8.7.6 Cell Suspension Culture and Secondary Metabolite Production

Nutritional, medicinal, and therapeutic content of various plant parts of *M. oleifera* has been reported repeatedly in several scientific journals, and Mahmud et al. (2014) provided a comprehensive metabolite database specific to each tissue of *M. oliefera*. This database provides details on metabolite spectra (using one-dimensional proton nuclear magnetic resonance) of different secondary metabolites and their biochemical pathways.

However, minimum studies have attempted to isolate desirable secondary metabolites from *M. oleifera* using cell suspension culture even though large numbers of studies have focused on its medicinal and nutritional values. Further studies should be concentrated on production of bioactive secondary metabolites using cell culture as it offers an opportunity to produce round the year production in a controlled environment which is not affected by climatic and soil conditions.

As an effective system of secondary metabolite production, cell suspension culture also ensures uniform quality, high quantity and purity, and continuous supply when compared with other isolation procedures. Such products have potential applications in medical industries as they are shown to possess anti-mutagenic, anti-carcinogenic, anti-aging properties.

However, the final yield of secondary metabolites by cell suspension culture through regular procedure is found to be low, and hence, use of certain enhancing procedures such as light irradiation with UV light, addition of jasmonic acid, ozone, heavy metal, ethylene, and sucrose seem to be increasing the yield of secondary metabolites.

Though industrial level production of food additives such as anthocyanins, shikonin compounds, safflower yellow, saffron, and other colorants have been reported using cell culture in other crops (Misawa 1994), incredibly rare reports are available in Moringa secondary metabolite production using plant cell suspension culture as detailed below.

As a first attempt, Mustafa et al. (2020) reported isolation of secondary metabolites using suspension cultures of Moringa tissues (especially leaves and seeds). They have also confirmed the specific antioxidant and antitumor activity of the isolated secondary metabolites against three cell lines, and highlighted that the Moringa callus extracts have presented higher antioxidant and cytotoxic activities than extracts obtained from the Moringa leaves and seeds as the Moringa calli were found to have higher amount of the secondary metabolites. This study provides the experimental proof that production of secondary metabolites from Moringa cell culture would be more productive than directly using Moringa leaf and seeds. In line with this, more experiments with Moringa cell suspension culture will shoot up soon as there are enough reports on importance of Moringa metabolites in therapeutic industries.

8.8 Genetic Engineering and Its Applications in Moringa

Crop improvement program generally utilizes introduction and selection, hybridization, and mutation strategies as the main methodologies.

However, transfer of desirable alleles through hybridization or increasing the genetic variation through mutagenesis requires several years of continuous work and huge resources.

Further, selection pressure on the breeding for desirable crop cultures for hundreds of years has also resulted in narrow genetic diversity in the cultivated crop gene pool and greatly hinders the further improvement in agronomically important traits. Though mutation breeding by employing both chemical and physical mutagens has been explored to expand genetic variation, it was found to be a random procedure and again this procedure is limited by its complicated and uncontrollable molecular mechanisms. Further, it is a labor- and time-intensive task as it needs to develop and screen huge numbers of mutants. Therefore, conventional breeding strategies seem to be exhausting, lengthy, and having untargeted results and cannot meet out the current demand on food, fodder, and fiber production. To this end, transgenic breeding or genetic engineering of crop plants is proposed as a viable strategy as it helps to evolve novel cultivars by transferring genes/traits from any exogenous biological system, which is not possible through conventional breeding.

Promises of genetic engineering or recombinant DNA (rDNA) technology towards increasing the agriculture productivity have been realized in several occasions (https://www.isaaa.org/kc/default.asp). Though it is not a standalone technology, but it has unique ability to transfer genes between entirely different organisms, and thus, it helps to ensure higher crop yields and enhanced nutritional value of the plant products. Actually, rDNA technology is considered as a complementary tool to traditional plant breeding approaches by facilitating fast track transmission of agronomically important traits among the crop plants.

Transgenic technology has revolutionized the plant breeding approaches, and globally 191.7 million hectares are planted with transgenic crops with improved traits (such as traits associated with herbicide resistance, pest and disease resistance, and quality) in 25 countries by 17 million farmers (www.isaaa.org accessed on 17thAugust, 2020). Another promising and more précise improvement in transgenic technology is *CRISPR/Cas* genome editing, which enable efficient targeted modification at nucleotide level. Recent developments in *CRISPR/Cas* systems and novel approaches in base editing may have great impact in breeding resilient crop cultivar as such developments will reduce the cost and skill required to develop efficient genome edited plants in several crops (Chen et al. 2019).

Despite such benefits shown in several economically important crops, transgenic technology has not yet been explored in Moringa. Transferring appropriate genes that control pest and diseases in Moringa would require immediate attention, as there are new pests (e.g., tea mosquito bug) and diseases emerging in some pockets of Moringa cultivation. At the same time, it should also be kept in mind that releases of genetically engineered organisms are not accepted everywhere in the world since it is speculated that they may have objectionable ecological, social, and economic effects. Further, lengthy and expensive regulatory process of evaluating transgenic crops and general acceptability of transgene(s) will limit the immediate commercialization of genetically modified Moringa.

References

Avila-Treviño JA, Muñoz-Alemán JM, Pérez-Molphe-Balch E, Rodríguez-Sahagún A, Morales-Domínguez JF (2017) In vitro propagation from bud and apex explants of Moringa oleifera and evaluation of the genetic stability with RAMP marker S Afr J Bot 108:149–155

Behera S, Kar SK, Rout KK, Barik DP, Panda PC, Naik SK (2019) Assessment of genetic and biochemical fidelity of fieldestablished Hedychium coronarium J. Koenig regenerated from axenic cotyledonary node on meta-topolin supplemented medium. Ind Crop Prod 134:203–213

Boopathi NM (2020) Forthcoming perspectives in MAS. In: Genetic mapping and marker assisted selection. Springer, Singapore, pp 481–504

Doyle J (1991) DNA protocols for plants. In Molecular techniques in taxonomy. Springer, Berlin, Heidelberg, pp 283–293

Gupta S, Kachhwaha S, Kothari SL, Jain R (2020) Synergistic effect of cytokinins and auxins enables mass clonal multiplication of drumstick tree (Moringa oleifera Lam.): a wonder. Vitro Cell Dev Biol Plant 1–12

Hassanein AM, Salem JM, Faheed FA, El-nagish A (2018) Effect of antiethylene compounds on isoenzyme patterns and genome stability during long term culture of Moringa oleifera. Plant Cell Tissue Organ Cult 132:201–212

Hassanein, A., Salem, J., Faheed, F., & El-Nagish, A. (2019). Some important aspects in Moringa oleifera Lam. micropropagation. Acta AgriculturaeSlovenica, 113(1).

Islam S, Jahan MAA, Khatun R (2005) In vitro regeneration and multiplication of year-round fruit bearing Moringa oleifera L. J Biol Sci 5:145–148

Mahmud I, Chowdhury K, Boroujerdi A (2014) Tissue-specific metabolic profile study of Moringa oleifera L. using nuclear magnetic resonance spectroscopy. Plant Tissue Cult 24(1):77. (Plant Tissue Culture & Biotechnology/Bangladesh Association for Plant Tissue Culture & Biotechnology; BAPTC&B)

Marfori EC (2010) Clonal micropropagation of Moringa oleifera L. Philipp Agric Sci 93(4)

Misawa M (1994) Plant tissue culture: an alternative for production of useful metabolites. FAO Agric Serv Bull 108:89

Mohan V, Purohit M, Srivastava PS (1995) *In vitro* micropropagation of Moringa pterygosperma. Phytomorphology 45:253–261

Murashige T, Skoog FA (1952) Revised medium for rapid growth and bioassays with tobacco tissue cultures. Physiol Plantarum 15:473–497

Mustafa R, El-Naggar EB, Svajdlenka E, Omran G, ELFiky F, El-Hawiet A (2020) Enhancement of phenolic content, antioxidant and cytotoxic activities of Moringa oleifera leaf and seed by suspension culture. Natl Prod Res 1–5

Nouman W, Siddiqui MT, Basra SMA, Afzal I, Rehman HU (2012) Enhancement of emergence potential and stand establishment of Moringa oleifera Lam. by seed priming. Turkish J Agric For 35(2):227–235

Ramachandran C, Peter KV, Gopalakrishnan PK (1980) Drumstick (Moringa oleifera): a multipurpose Indian vegetable. Econ Bot 275–283

Saini RK, Shetty NP, Giridhar P, Ravishankar GA (2012) Rapid in vitro regeneration method for Moringa oleifera and performance evaluation of field grown nutritionally enriched tissue cultured plants. 3 Biotech 2(3):187–192

Salem JM (2015) In vitro propagation of Moringa oleifera L. under salinity and ventilation conditions. Genet Plant Physiol 5(1–2):54–54

Steinitz B, Tabib Y, Gaba V, Gefen T, Vaknin Y (2009) Vegetative micro-cloning to sustain biodiversity of threatened Moringa species. Vitro Cell Dev Biol Plant 45(1):55

Stephenson KK, Fahey JW (2004) Development of tissue culture methods for the rescue and propagation of endangered Moringa spp. germplasm. Econ Bot 58:115–124

Xiang SQ, Liang GL, Guo QG, Li XL, He Q (2007) Tissue culture and tetraploid induction of drumstick. J Trop Subtrop Bot 02

Molecular Markers and DNA Barcoding in Moringa

9

N. Manikanda Boopathi and M. Raveendran

Abstract

Evolving novel crop cultivars with improved agronomic performance and increased yield and quality traits using traditional breeding approaches is a time-consuming, labor intensive, and elaborate process. Recent developments in molecular marker technologies and sophisticated tools in genetic mapping and quantitative trait loci localization in the plant genome, offered unique and promising avenues in molecular breeding. It speeds up the routine breeding process by precisely introgressing and evaluating the breeding materials. Since the first introduction of restriction fragment length polymorphism (RFLP), several types of molecular makers (mostly random DNA markers, which are phenotypically neutral and intricately linked to the gene of interest) have been developed and successfully employed in crop improvement program during the past few decades. An overview of different kinds of molecular markers and their applications is provided in this chapter, besides providing a snapshot on the molecular markers and its applications in Moringa. This chapter also provides particulars on basics of DNA barcoding, genes that are suitable for Moringa DNA barcoding, and its applications in Moringa trade.

N. M. Boopathi (✉) · M. Raveendran
Department of Plant Biotechnology, Centre for Plant Molecular Biology & Biotechnology, Tamil Nadu Agricultural University, Coimbatore, India
e-mail: nmboopathi@tnau.ac.in

9.1 Prelude on Molecular Markers

Evolving elite cultivars in crop plants including Moringa has been attempted using different conventional breeding strategies such as introduction, selection, hybridization, mutagenesis, development of synthetic and composite varieties. In fact, all the cultivars released in Moringa have been developed using the conventional breeding methods (see Chap. 5), which involves evaluation of traits (such as pod yield and quality) under natural conditions (famer's field) and/or greenhouse conditions and selecting the best performing lines. However, such selection process is highly influenced by the soil and other input resources and prevailing climatic conditions.

Recent advances in molecular markers and genetic maps have shown their potentials in fast and accurate recovery of the target traits in several field crops, and hence, it is anticipated that employing marker technology will also speedup the Moringa crop improvement process. Use of markers in breeding program has long history and it started with the employment of morphological traits (such as seed shape and color) as first-generation markers. However, owing to their limited number and adverse effects of

N. M. Boopathi et al. (eds.), *The Moringa Genome*, Compendium of Plant Genomes,
https://doi.org/10.1007/978-3-030-80956-0_9

environment factors on expression of those traits, isozymes or biochemical markers were alternatively used as a second-generation markers to indirectly select the plant breeding materials. Soon, it was comprehended that isozymes were also not supportive for such breeding efforts owing to the facts that they are also limited in number, unstable during experimental assays, and possessing poor genetic relationship with the target traits (i.e., all the agronomically important traits are not always related with appropriate isozymes). As research intensified in deoxyribose nucleic acids (DNA) and ribo nucleic acids (RNA), the third-generation markers viz., DNA and RNA markers became popular in plant breeding, and they almost replaced the previous generation markers.

The key reasons for employing third-generation markers in plant breeding are:

(1) Identification of molecular markers are not involved in any pleiotropic and epistatic effect on the phenotype and they are unbiased in phenotypic reactions.
(2) Segregation and inheritance of DNA markers are not affected by the environment.
(3) Assays that used to capture DNA marker data are robust and can be automated.
(4) DNA or molecular markers are stable and detectable in all parts of plants irrespective of their growth and development and not affected by any factors.

Though there are different types of molecular markers, in order to realize their full potentials in plant breeding program, a perfect molecular marker for the plant breeding program should have the following characteristics:

(1) It should generate highest level of polymorphism even between closely related plant species (which are frequently used in plant breeding program due to their compatibility during hybridization) and be evenly distributed throughout the genome.
(2) The marker should be independent and reliable and should generate multiple alleles.
(3) The experiment used to assay the selected marker should be simple, swift, and cheap.
(4) The marker assay should require little tissue and/or DNA samples.
(5) Development of the marker assay should not depend on genomic information of an organism.
(6) The identified polymorphic marker should be linked to large numbers of agronomic- and yield-related traits.
(7) It should offer adequate resolution (minimum distance) to the identified quantitative trait loci (QTL; see Sect. 9.1.2).

On the other hand, there is no molecular marker that can be used as an ideal marker as each marker class has its own advantages and disadvantages (Table 9.1). Assay that used to detect each marker class as well as their applications largely differ from each other at their abundance in the given genome, polymorphic information content between the investigated parental lines, specific ability to detect the target QTLs, reproducibility across the laboratories, and ease of the assay and their affordability. Depending on the need, several modifications in the assay techniques can be made, leading to a next generation of advanced molecular markers.

9.1.1 Types of DNA-Based Markers

Specific fragment of the DNA/RNA that has the ability to distinguish two or more individuals of the crop plants is called as a molecular marker. Usually, in a typical gel-based assay, all the molecular markers produce a band-like structure in the lane that belong to each accessions, and the banding pattern (which is generated by the length differences of alleles of the each accessions; Fig. 9.1) is used to differentiate the individuals. The publication of Botstein and his team in 1980 on restriction fragment length polymorphism (RFLP) was the first report that described the molecular marker and its utilization in the construction of genetic maps.

For simple understanding, marker techniques are grouped into two categories: (1) non-PCR-based or hybridization-based techniques (e.g.,

Table 9.1 Summary of applications and restrictions of different classes of molecular markers when they are employed in plant breeding program

Type of marker[a]	Advantage	Limitations
Protein markers	1. Rated as technically simpler and lesser expensive markers when compared with other marker classes 2. Markers are codominantly inherited	1. Proteins are unstable and it is difficult to differentiate the assay results and experimental errors 2. As less than 30% of the nucleotide variations lead to changes in protein mass (and hence changes in protein fragment sizes), majority of the isozymes cannot be detected as polymorphic marker, as they result into similar banding pattern 3. Further, the polyploidy nature of majority of the cultivated plants interfere with the interpretation of allozymes patterns
RFLP	1. Extremely specific to the given locus 2. Inherited in true Mendelian fashion 3. Show highest reproducibility among the marker classes. 4. Shown to possess adequate genomic abundance	1. Large quantity of pure DNA is required for the assay 2. Need prior genomic sequence information to develop probe for the assay 3. Involves radioactive and toxic reagents in the assay 4. Require skilled manpower and cannot be automated
AFLP	1. Among the marker classes, shown to generate highest number of Multi-locus markers (can produce 50–100 polymorphic fragments between closely related genotypes per assay)	1. Though it can be semi-automated, needs heavy investment and elaborate procedure, and 2. Require skilled workforce
RAPD	1. Need minimum laboratory infrastructure 2. Identified as simplest assay technique 3. As a startup in genome research, it can be applied to any genome	1. Reproducibility across the laboratory is the frequent problem associated with this marker due to the fact that a single variation in the experimental procedure will lead to produce errors. However, multiple replications to confirm the polymorphism will avoid this limitation 2. RAPD markers are dominantly inherited
SSR	1. Simple assay technique with minimum requirement of infrastructure 2. As the markers are codominantly inherited, it is easy to identify the heterozygotes from the homozygote parents 3. Regarded as the breeder's marker of choice due to their extensive effectiveness	1. Largely depends on the repeat motif identification, which require prior genomic information of the given species 2. SSR have relatively poor discriminative power among the closely related accessions
SNP	1. Automated and large-scale genotyping assay methods assist in the detection of highest numbers of informative markers 2. Variations in the coding regions may or may not change the amino acid sequence, however, it can be effectively linked to the trait of interest 3. SNPs are employed in fine mapping by developing high-density genetic maps which is useful for positional cloning (a process by which the underlying gene is cloned and sequenced)	1. Largely depend on the genomic information 2. Require relatively higher investment for initial infrastructure development

[a]See Sect. 9.1.1 for expansion of each marker class

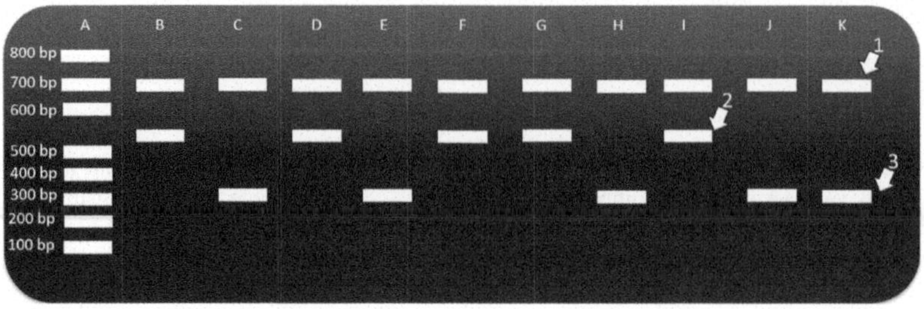

Fig. 9.1 A schematic illustration of molecular markers and its usage in differentiating the genetic variation among the plant accessions. A—marker ladder lane, which is used to denote the approximate size of the DNA fragments that moves in the gel. Lanes B to K represents different Moringa ecotypes. Arrow heads with numerals denotes different alleles produced by the same marker (say for example, RAPD) that can be able to differentiate the ecotypes. In this illustration, Allele 1, is common for all the ecotypes and hence said to be monomorphic. Whereas, Allele 2 and Allele 3 are said to be polymorphic as they have different banding pattern among the Moringa ecotypes. Allele 2 can be used to differentiate the ecotypes B, D, E, G, and I from the investigated ecotypes B–K. Allele 3 can be used to differentiate the ecotypes C, E, H, J, and K from the investigated ecotypes B–K

RFLP) and (2) PCR-based techniques (e.g., SSR and RAPD).

Novel molecular marker systems were developed after the invention of polymerase chain reaction (PCR) technology due to its minimalism in experimental procedure and superior chance of getting polymorphism. Another major breakthrough is the use of synthetic random primers for PCR amplification (and hence such marker class development does not require prior sequence information) for PCR analysis which facilitated the introduction of various types of molecular markers for a range of applications in crop improvement.

PCR-based marker development are further grouped into two subcategories: (1) arbitrarily primed PCR-based techniques or randomly targeted sequence amplification techniques (e.g., Random amplified polymorphic DNA (RAPD), Amplified fragment length polymorphism (AFLP)), and (2) specific sequence-focused PCR techniques (such as microsatellite (or short tandem repeats or simple sequences repeats, SSR) based marker technique, Inter Simple Sequence Repeats (ISSRs), Randomly Amplified Microsatellite Polymorphisms (RAMP; which is a combination of RAPD and ISSR), Cleaved Amplified Polymorphic Sequences (CAPS), Sequence Characterized Amplified Regions (SCAR), Sequence-Related Amplified Polymorphism (SRAP), Target Region Amplification Polymorphism (TRAP), Single Strand Conformation Polymorphism (SSCP), Inter-Retrotransposon Amplified Polymorphism (IRAP) and REtrotransposon-Microsatellite Amplified Polymorphism (REMAP), Retrotransposon-Based Insertion Polymorphism (RBIP), Single Nucleotide Polymorphism (SNP)).

As science advances in genomics and transcriptomics, those data have also been employed to identify novel marker classes. For example, molecular markers were developed by PCR amplification of expressed region of the genome (exons) and they are defined as genic molecular markers. Such markers are functional portions of the given genome and they are mainly developed by using the complementary DNA or Expressed Sequence Tags (cDNA/EST) datasets. Further, huge amount of small RNA sequences have also been generated due to the rapid progress in next-generation DNA sequencing which led to develop microRNA-specific novel functional markers at the DNA level. Detailed discussion on the marker assays for different kinds of molecular markers and troubleshooting measures can be found at Boopathi (2020).

The important inherent properties of molecular markers (Tanksley 1983) that distinguish them from morphological markers are:

1. Morphological markers required whole plant assay whereas molecular markers can be determined at tissue or cellular levels.
2. A relatively lesser number of informative morphological markers are available (which also require application of exogenous mutagens); however, in nature, abundant informative alleles are available at the molecular level.
3. It has also been noticed that the given morphological marker very often linked with undesirable phenotypes. In contrast, alternate alleles of molecular markers rarely reported to be associated with deleterious effects.
4. Majority of the molecular markers segregate codominantly, and hence, both homozygous and heterozygous alleles can be distinguished easily; whereas dominant segregation of morphological markers limit their utility in many breeding applications.
5. Unlimited numbers of molecular markers with fewer epistatic or pleiotropic effects can be explicitly scored in the segregating generations, which is not at all possible with morphological markers.

9.1.2 Applications of Molecular Markers

Applications of molecular markers have been realized in analyzing genetic diversity, identifying molecular phylogeny and evolutionary relationships, detecting duplications in germplasm collections, identifying close relatives, fingerprinting varieties/cultivars/hybrids, mapping genetic loci and tagging genes for management and improvement of crops, conserving endangered germplasm species, discriminating male and female flowers, etc.

Codominant molecular markers were primarily employed to fix the hybrids. Dominant markers cannot distinguish the allelic difference of a gene when it is in heterozygous conditions whereas codominant marker can differentiate the allelic difference. Therefore, codominant markers are used to categorize the hybrids simply and effortlessly from their homozygous parents. The markers coming under the codominant category are protein markers (such as Allozymes and Isozymes) and DNA-based single locus markers (such as RFLP, SSR, and SNPs). RAPD, ISSR, and AFLP are called as dominant markers.

Markers with increased reliability and functional information can be developed using bioinformatic tools when the sequences from the genic regions are used. As these markers are developed from genic regions, any polymorphism detected in the individuals, varying for the given trait, will facilitate in exploring the underlying biological function. Besides, such markers can be used as potential candidates for comparative genomic studies and offer an accurate estimate of functional diversity. The data developed from these studies would be useful to understand the molecular mechanisms involved in acclimatization of plant species to diverse environmental conditions.

Large variations in reproductive systems such as hermaphrodite, monoecious, and dioecious have been recognized in flowering plants. Identification of gender-specific plants, at the seedling stage itself, will significantly reduce the time, effort, and resources required to sustain the female plants in the field, to guarantee greater yields. Hence, sex-specific molecular markers have shown its applications in the crop breeding program with greater precision.

In view of its immediate potential applications of molecular markers in Moringa, only two applications are described in detail in the below subsections.

9.1.2.1 Marker-Assisted Selection

Molecular markers have shown to play different roles in plant breeding program, and the immediate application of DNA-based molecular markers in Moringa is developing high-resolution genetic (or linkage) maps and detecting molecular markers tightly linked with genes or quantitative trait loci (QTLs; which comprises a set of genes) underlying agriculturally important quantitative traits (such as disease and insect resistance, abiotic stress tolerance and pod or leaf yield, and other economic quality related characteristics). Comprehensive procedures for constructing linkage map and analysis of QTLs in plants have been described by Boopathi (2020). Linkage mapping and QTL analysis have provided opportunities for indirect selection of economically important traits using the linked markers instead of the trait itself. Such selection process is known as marker-assisted selection (MAS).

MAS has been considered to be superior over desirable trait-based selection due to the following advantages:

i. Rapid and simple selection of the desirable traits even without the involvement of selection environment (e.g., for the selection of certain disease resistance traits, MAS is considered to be quicker, cheaper, and more efficient than the routine phenotype-based selection; besides selection can be performed even when there is no disease incidence).
ii. Some of the economically important traits are expressed and can be scored at the end of the growing season, and further, they are regulated by the unstable developmental process in response to the environmental conditions. Use of molecular markers linked to the trait enables early selection for such traits and breeder's convenience.
iii. Molecular markers also facilitate selection of multiple traits from the same plant population that are grown in the same season and environment. Such effort is not possible through conventional breeding. For example, selection of drought and submergence tolerance traits requires evaluation of plant population under different seasons and environments. On the other hand, plants with drought and submergence tolerance can be selected using the respective markers linked to these traits without conducting the field evaluation trials.
iv. Without performing any progeny testing for target traits such as disease resistance, selection for recessive traits can be done with molecular markers.
v. Use of foreground, background, and recombinant markers in MAS enables rapid removal of undesirable genetic backgrounds as well as plant materials with undesirable gene combinations in segregating populations during the early stage of plant breeding program, and thus, greatly reduces the breeding cycle besides retaining the most useful plant materials.
vi. Selection of progenies using molecular markers greatly reduces the problems associated with linkage drag, i.e., it efficiently breaks the linkage between favorable and unfavorable alleles.
vii. The traits that show low heritability and exhibit high genotype x environment interactions (and hence such traits cannot be selected efficiently due to environmental effect during conventional selection), can be selected effortlessly using molecular markers.
viii. As the MAS is independent of seasons and environments, multiple numbers of selections can be done in a year. Thus, MAS has more pronounced applications in speed breeding.

Owing to the above advantages, molecular markers are currently being employed extensively in both public and private sector crop breeding programs to genetically improve the target traits in the given crop.

On the other side, although molecular markers have been identified and found to be associated with several agriculturally important traits, they are not always consistent or having their immediate application in plant breeding programs. Main confronts in assemblage of markers to the trait of interest for efficient MAS is identifying markers that are closely linked, inexpensive, and simply assayed. However, majority of the identified markers for MAS are unsuccessful in proving their efficiency as these identified markers are not either tightly close to or not part of the gene itself. Even tiny genetic distance identified between the markers and the genes of interest results in genetic recombination, and ultimately, leads to loss of the trait of interest.

Exchange of genetic elements between the marker and the gene may occur through the process known as genetic recombination during the evolution or plant breeding process. Owing to this genetic recombination, even though the plants are selected with the associated markers, selected plants may not have the desirable gene in the favorable allele combinations. Such plants are called as false positives. Apparently, such increase in false positives among the progenies will lead to decrease in the breeding efficiency. Therefore, validation of identified molecular markers (by confirming its closeness to the target gene either by fine mapping or positional cloning) before implementing the MAS will greatly increase the utility of the given molecular marker in crop improvement programs.

Next difficulty in widespread utilization of MAS in crop improvement programs is the need for identifying consistent marker alleles linked to the target genes across the germplasm and environments. In practice, MAS has been found to be effective when the markers are linked to the trait (s) of interest across the breeding populations. Unfortunately, in majority of the reports, the identified markers were not validated across the global (or at least using the national) breeding populations.

Thus, cautious development and validation of markers are required before implementing them as routine indirect selection criteria. Though the use of molecular markers has shown its widespread applications in field and horticultural crops (Boopathi 2020), their potential applications have not yet been fully realized in Moringa. Those scientific reports that focused on use of molecular markers in Moringa are summarized here under.

9.1.2.2 Documenting Genetic Variation in Moringa Using Molecular Markers

Understanding the molecular mechanisms of different biological process, growth, and development can be revealed by carefully designed experiments that explore the plant genetic variation at molecular level. Though the next-generation sequencing strategies provide the opportunity to sequence the plant genome relatively with simple efforts, there is still some issues in getting the complete plant genome; besides, it would be heavily priced, if all the associated plant kingdom is targeted to sequence. To this end, molecular markers would be an cost-effective and simple approach not only to capture the plant genetic variation by offering necessary genomic landmarks in the given plant species, but also it provides methods to correlate the phenotypes with the identified polymorphic markers.

Studies that focused on ecological, evolutionary, taxonomical, phylogenic, and genetic experiments in plant sciences have been successfully and now being routinely conducted with molecular markers such as RFLP, RAPD, SSR, and AFLP. Their utilities as well as restrictions associated with these markers in unraveling the plant genetic variation have been realized (Boopathi 2020).

Using the basic principles of early generation markers development, advanced and semi-automated techniques (primarily based on the principle of detecting the SNPs) have emerged and being employed to high-throughput analysis of the genetic variation with large numbers of samples in a short span of time. Such advanced methodologies (that detect the variations in the form of SNPs, structural variations (SVs), Insertions and Deletions (InDels)) have evolved to combine useful features of previous marker techniques and also increased the sensitivity and resolution to detect genetic discontinuity and distinctiveness. Besides, the recent marker development also focuses to capture the variations in other untapped genomic regions such as retrotransposons, mitochondrial, and chloroplast-based microsatellites. Such efforts increased the genome sampling exposure and thus provided extended information on genetic variation. However, it involves initial investment in sequencing the organelle genomes and targeted genomic repeat rich regions in the given plant germplasm.

Alternatively, earlier RAPD and AFLP principles are employed but using cDNA library specifically developed from the given plant tissues that are under various development stages or stressed by different biotic and abiotic factors. Such efforts greatly increase our ability to study different types of gene expression models and reveal molecular mechanisms behind such biological responses which would have practical utility in designing precise plant breeding agenda.

9.2 Molecular Markers in Moringa

The genomic information on Moringa was reported as early as 1937 by mentioning "*Moringa tree is a true diploid with 2n = 28*" (Patel and Narayana 1937). However, complete potentials of molecular markers have not yet been realized in Moringa, and they are mainly explored extensively to document the genetic diversity that exist in the different Moringa germplasm accessions that have been maintained at different institutes.

Among the different classes of markers, majority of the studies employed randomly generated multi-locus markers such as AFLP, RAPD, and ISSRs as these are found to provide significant variation among the investigated Moringa accessions and invariably established that these marker classes are rapid and cost-effective fingerprinting techniques (Table 9.2). For example, Hassanien and Al-Soqeer (2018) assessed the genetic diversity of *Moringa oleifera* and *Moringa peregrina* that were present in Saudi Arabia using ISSR markers, whereas Kumar et al. (2017) found RAPD markers were useful to capture the genetic variations and explore the genetic relationships among seven advanced breeding lines of Moringa.

Another striking point from the published reports is that there is a significant genetic variation existing even among the investigated closely related Moringa accessions. This may be due to out-crossing that occurs frequently in Moringa (Actually, it is a cross-pollinated crop). This implies that those genetically diverse unique accessions would be a valuable resource for further breeding program that focus on novel traits, such as improved nutritional content, and conservation of such genetic variation for the future.

9.2.1 SSR Marker Development in Moringa

The need for development of SSRs in Moringa has been first described in Wu et al. (2010), and they amplified microsatellite loci using the customized enrichment protocol that employed AFLP of nucleotide repeats containing sequences (Zane et al. 2002). As per the protocol, Wu et al. (2010) screened 288 randomly selected clones by colony PCR and identified that 210 of them were microsatellite motifs. Though 209 of these clones were productively sequenced, only 192 clones (67%) were possessing SSRs. Consequently, primers were designed to flank these microsatellites and found that out of 69 pairs of 20-base primers, only 46 pairs successfully amplified target regions. Finally, they reported 20 Moringa SSRs that can offer useful polymorphism among Moringa germplasm. It can be noticed from this study that the strategy employed for development of novel SSRs was a laborious and time-consuming process besides yielding a poor number of polymorphic SSR markers.

Alternatively, to increase the efficiency of reporting large numbers of novel SSRs, various molecular strategies were employed. For example, selective hybridization of microsatellite regions of the genomic libraries was regarded as

Table 9.2 Selected studies that focused to utilize molecular markers for the assessment of genetic variation among the Moringa genotypes

Investigated plant materials	Markers employed	Remarks	References
This study utilized seven different populations (two from Tamil Nadu (South India), one from Ex-Nsanje region (Southern Malawi), and four from Kenya with 20 replications (totally 140 genotypes)	AFLP	If the *M. oleifera* genetic resources were explored from the regions in which they have originated, they show increased chance of documenting novel genetic variation. AFLP can be used to capture maximum number of informative alleles among the marker classes	Muluvi et al. (1999)
Sixteen accessions from Germplasm Bank (BAG) of Embrapa Coastal Tablelands, Sergipe, Brazil	RAPD	Owing to low diversity detected in this investigation, new activities of collection was proposed to increase the diversity	da Silva et al. (2012)
Eight Moringa cultivars of South India	Both random and gene-specific markers (RAPD, ISSR, and cytochrome P_{450}-specific)	Though the investigated cultivars were classified into four sub-clusters, there was no geographical-based grouping. This confirmed active spread of the planting materials. Further, as Moringa is a cross-pollinated crop, there was great rate of gene flow	Saini et al. (2013)
Totally 12 populations comprising 300 genotypes assembled from northern (Himachal Pradesh) and southern (Tamil Nadu) states of India	SSR	SSR-based grouping disagree with geographical distribution. Hence, concluded that there was seed movement and/or high rates of gene flow between the adjacent populations	Ganesan et al. (2014)
Seven advanced breeding lines of Moringa	Different types of markers such as seed protein profile, RAPD, and Cytochrome P450 gene-specific markers	Only RAPD and Cytochrome P450 markers generated considerable amount of polymorphism, 79.68 and 86.44%, respectively	Kumar et al. (2017)
Ninety-seven Moringa accessions assembled from various parts of India	SSRs	Reported high genetic diversity among the investigated accessions	Rajalakshmi et al. (2017)
Tissue cultured Moringa plants	RAMP	RAMP can be used to ensure genetic stability of the different systems of micropropagation of *M. oleifera*	Avila-Treviño et al. (2017)
Moringa oleifera and *M. peregrina*	Morphological markers and ISSRs	ISSR were effective in the characterization of genetic diversity	Hassanien and Al-Soqeer (2018)

(continued)

Table 9.2 (continued)

Investigated plant materials	Markers employed	Remarks	References
Seven genotypes per species of *Moringa oleifera* and *M. peregrina*	Nutritional characteristics and SSR	A strong relationship was found between nutritional and molecular classifications of genotypes	Hassanein (2018)
Ninety-seven *M. oleifera* accessions (2 hybrids and 95 naturally grown traditional varieties)	ISSR and SRAP	These markers are useful for population structure and association studies	Rajalakshmi et al. (2019)
Leaf samples collected from 10 M. *oleifera* plants that were naturally found in Middle Delta, Egypt	Random markers such as ISSR, start codon targeted (SCoT) markers, and barcoding genes such as *ITS2*	ISSR, SCoT, and ITS markers can be employed to effectively identify *M. oleifera* products from its adulterants and similar species	Hassan et al. (2020)
Totally, 23 M. *oleifera* accessions that were collected from Kerala, Tamil Nadu, and Karnataka states of India	Seven different cytochrome P_{450} ($CytP_{450}$) gene-specific markers	$CytP_{450}$ can be used to capture genetic diversity in Moringa	Ravi et al. (2020)

robust, high-throughput, reproducible, and inexpensive method. The steps involved in selective hybridization facilitate securing large-scale microsatellite sequences that are labeled with biotin probes and consequently capturing them using magnetic beads coated with streptavidin. Alternatively, the probes can also be fixed on nitrate filter. This kind of enrichment of SSRs and their isolation have been successfully employed in different crop species including Guava. However, this method is again found to be involved in high demand for infrastructure facilities, skilled manpower, and additional time.

Another novel strategy proposed for large-scale development of SSR is use of next-generation sequencing data for identifying SSR motif and designing primers that flank such motifs. Thanks to the widespread availability of next-generation sequencing facilities, nowadays, draft or complete genome sequences and transcriptomic sequences of several crop species including Moringa (refer Chap. 10 for more details) are available for public use.

Presence of microsatellite markers in those sequences can be easily identified by using several algorithms (for example, the MIcroSAtellite (MISA) identification tool (http://pgrc.ipk-gatersleben.de/misa/misa.html); SSRIT available in Gramene (http://gramene.org/ssrit) and SSR-Locator (www.microsatellite.org/ssr.php)). Such software offer wide range of parameters that can facilitate the rapid and effortless identification of perfect di-, tri-, tetra-, penta-, and hexanucleotide motifs with the customized minimum number of repetitions. While identifying the SSRs, mononucleotide repeats usually are not considered as it may lead to misinterpretation of results.

Subsequently, the primer pairs will be designed by using software such as Primer3 v0.4.0 with a set of adjustable parameters. The following widely employed parameters would be useful to design robust primer pairs: primer size 18–25 bp (optimum 20 bp), PCR product size 100–400 bp (optimum 280 bp), GC percent 45–70 (optimum 50%), Tm 57–63 °C (optimum 60 °C).

The SSRs identified in the above workflow must be validated with the target crop, and it can also be tested among the related species to test verify their transferability among the closely related species. More than thirty thousand such genome-specific and genic SSR markers (Fig. 9.2) were developed at this laboratory using the genomic and transcriptomic sequences of Moringa cultivar PKM1 (article in preparation) for the first time. A set of randomly selected

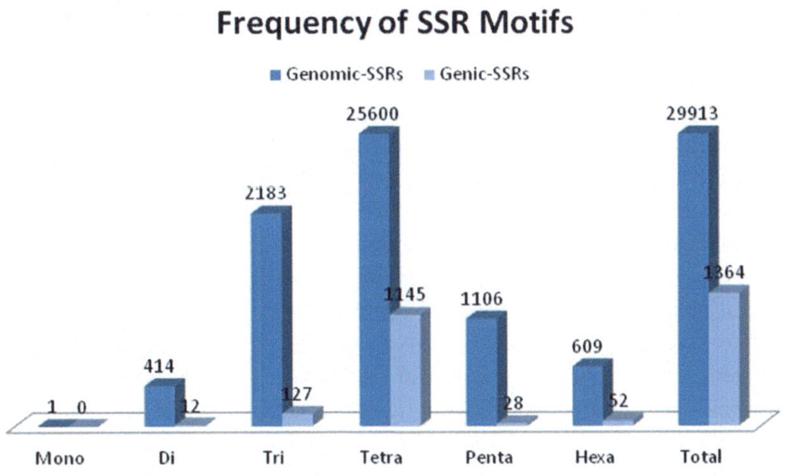

Fig. 9.2 Summary of SSRs identified using resequencing and transcriptomics sequence of Moringa cultivar PKM1

SSRs were validated in Moringa ecotypes and proved their utility in Moringa research. Informative SSR markers identified in this exercise will have several applications such as varietal protection, germplasm management, trait mapping, gene isolation and selections, biotic and/or abiotic stress resistance improvement using molecular breeding strategies.

9.3 Prelude on DNA Barcoding

DNA barcoding is the procedure for identification of crop/animal/microbial species using the variations at nucleotide level that are found in the specific DNA fragments (DNA barcodes). DNA barcoding has been first established in animals using *cytochrome c oxidase subunit 1* (*COI*) gene, and now, it is being used as a standard DNA barcode in animals.

However, owing to the lack of universally accepted DNA barcoding gene(s), it is yet to be established in plants. Use of several plant candidate genes from plastid and nuclear genome, such as *atpF-atpH* spacer, *matK*, *rbcL*, *rpoB*, *rpoC1*, *psbK-psbl* spacer, and *trnH-psbA* spacer, have resulted in that there is no single gene that can be used for DNA barcode in plants. Consequently, Consortium for the Barcode of Life (CBoL)–Plant Working Group recommended a two-gene combination, *rbcL* + *matK,* as the plant DNA barcodes by comparing the sequence quality, reproducibility across the laboratories, relative efficacy of testing, and levels of species discrimination.

Employment of *rbcL* + *matK* as DNA barcodes has been shown to be universal framework in DNA barcoding of land plants and identify and discriminate the closely related species and ecotypes within the species. It is recommended to initiate DNA barcoding of any land plant species with this two-locus standard barcode (*rbcL* + *matk*) as it has reported so far a minimum of 70% species discriminatory power; however, additional DNA barcoding genes need to be explored if superior resolution is required.

Retrieval of better intraspecific than interspecific variation is considered as the ideal qualities of a DNA barcoding region in such a way that highly similar species can be recognized using those unique variations in barcodes. Additional qualities of an ideal DNA barcoding region includes they should be short (~750 bp), universally amplifiable across all plant taxa using simple procedure, generated sequences should align readily with the reference sequences, and represents low copies of insertion-deletions (InDels).

Key application note of DNA barcoding is to provide a common community resource of DNA sequences which can precisely identify the organisms and clarify the taxonomical classifications. Such barcoding genomic resource shed light on species-level taxonomy among the individuals having simple and similar morphologies, those with exaggeratedly broad classifications, those that are tiny in size (or powdered for easy of packing) and cannot be handled at morphological level, and those that are inadequately characterized even though they possess genetic diversity due to the complex procedure of morphology-based taxonomy.

DNA barcoding is now widespread among several areas of plant sciences and professionals such as taxonomists, ecologists, conservationists, foresters, agriculturalists, forensic scientists, customs, and quarantine officers as they are routinely employing DNA barcoding for plant characterization. Hollingsworth et al. (2011) have reported that discrimination power of the given DNA barcode can be very much useful and the following are the examples of such instances:

(1) Characterization of genetic diversity of plant species exist in the given geographical regions where immature material and/or plant fragments need to be identified.
(2) Identification of the plant species in the processed and packed plant products (such as diet supplements prepared from herbs, food products, and the components of herbal medicines). Usually, the challenge is to identify the adulteration of different or related species in the target plant product and notify the members of other taxonomic groups rather than given species.
(3) Classifying the unfamiliar plant species with the existing species group when there is no information on the investigated plants. It is also found to be useful in classifying the species rich plant systems where there is poor availability of expertise in taxonomy.
(4) Application in ecological forensics to critically characterize plant roots, seedlings, or cryptic life stages (e.g., fern gametophytes) where DNA barcoding can be used as an affordable practical route to precisely identify the specimens. For example, *rbcL* sequences alone was successfully employed by Kesanakurthi et al. (2011) to characterize 85% of all root samples and provided a detailed report on ecological factors that played important role in the subterranean spatial organization of plant diversity. Similarly, Harmon (2010) used *matK* DNA barcodes to notify unspecific plant species in herbal supplements, and found that herbal supplements of Black Cohosh did not contain the target north American species *Actea raceomosa* and instead it was a mix of Asian species of *Actea*.
(5) Other forensic studies in which DNA barcoding was employed are (i) predicting poached wildlife, (ii) pinpointing forensic relevant fly species in forensic cases, (iii) recognizing species in illegal egg and timber smuggle as well as in ecological forensic studies, (iv) monitoring biological invasions in soil as well as water, and (v) stressing the need for upgradation of taxonomic revision.
(6) Protecting the endangered species in trade is also effectively mediated by DNA barcoding. Convention on International Trade in Endangered Species of Wild Fauna and Flora (CITES; http://www.cites.org/eng/disc/species.shtml) has protected more than 29,000 plant species, and it is important to find effective and affordable methods to discriminate CITES-listed plant species from non-CITES-listed plants. To this end, *matK* DNA barcodes were found to be efficient to differentiate traded timber products of *Ramin* (*Gonostylus*) species which are CITES protected (Ogden et al. 2008).

Though DNA barcoding is proved useful in initial identification and categorization of plants at species level, a major limitation of DNA barcoding is it lacks precise definitiveness of the plant species (as that of animals) that can be accepted by the court of law and other regulatory frameworks. In order to realize the full potential of DNA barcoding, it is vitally required to develop an appropriate reference library which warrants production of a database on DNA barcodes from vouchered plant samples that were completely characterized. Several such geographically- and taxon-based projects are being undertaken in the development of DNA barcode database including in an Indian Moringa cultivar,

PKM1 (see below). Besides, sequence of different DNA barcoding genes from different plant species is also stored in GenBank, which could be a valuable and useful resource for taxonomic classifications at molecular level.

However, those data were collected and archived by the individual scientists, and it lacks validation with the voucher information and linking with the passport data of the specimens. This ultimately leads to a difficulty in identifying and removing wrongly identified specimens or irrelevant sequences. Therefore, it is imperative to link the vouchers, sequences, trace files, and other metadata by following BARCODE standards, as it is done in the development of GenBank databases and Barcode of Life Data systems (BOLD).

Success of herbal drug industry chiefly depends on precise, rapid, and scientifically authenticated identification of the plant products. Conventionally, it has been done with a taxonomic expert, who employs combinations of both traditional and modern tools (such as morphological characteristics and biochemistry-based methods) to identify the plant materials. However, such approach would be elaborate, lengthy, and costlier for bulk identification of plant materials at industrial scale. Besides, trends in high processing of plant products with numerous ingredients may not offer accurate result with the conventional methods. This necessitates speedy, accurate, and affordable tools to screen the huge volume of plant products.

9.4 DNA Barcoding in Moringa

Even though India has an enormous genetic wealth of *Moringa oleifera* found in varied geographical locations, attempts towards capturing such variations in its fullest potential are limited. Existence of variability in this crop is not only because of adaptation to varied environmental conditions, but also due to the mechanism of cross-pollination. Therefore, such wide genetic variations and hidden special genetic characteristics are yet to be fully explored to document their uniqueness. Though there are several attempts in DNA barcoding of large numbers of agricultural and horticultural crops, it has not yet been explored in Moringa.

Besides documenting genetic variation and its potentials in crop improvement programs in Moringa, DNA Barcoding can also be utilized in authenticating the Moringa leaf products in both regional and global markets. Recent past decades have witnessed restoration of health care with plant-based products, and consequently, there has been an elevated demand for herbal products and drugs in global trade. Particularly, Moringa become an important export and import food commodity as it has been considered as the natural nutrition of the tropics.

9.4.1 Selection of Appropriate DNA Barcoding Genes

As outlined above, quality control of Moringa supply systems can be effectively managed with the help of DNA barcoding. However, application of DNA barcoding in Moringa critically require seamless protocols for DNA barcoding and fool-proof method of DNA sequence-based identification that are affordable to industry and officially agreed by the competent authorities for their role in regulated procedures. DNA barcoding genes that can be used for herbal plant identification includes *matK, rbcL, trnH-psbA, ITS, trnL-F, 5S-rRNA,* and *18S-rRNA*.

As highlighted in the Sect. 9.3, in order to select a universal DNA barcode gene for plants, the Consortium for the Barcode of Life (CBOL) initiated the formation of a working group and tested the utility of seven leading candidate barcoding genes on the following parameters: universality (ease of amplification and sequencing), sequence quality, and discriminatory power. However, this project finally concluded that there is no single gene that can be used for all the plant species as each gene has its own strengths and weaknesses. Consequently, CBOL Plant Working group suggested two plastid coding regions, *rbcL* and *matK*, to initiate the DNA barcoding in plants and when it is required, it can be complemented with additional genes such as *ITS* (Hollingsworth et al. 2011).

The reference for *rbcL* barcode was obtained from complete *Arabidopsis thaliana* plastid genome sequence (gi 7,525,012:54,958–56,397) which targets a 599 bp region at the 5' end of the gene (starting at 1 bp and ends at 599 bp, including primer sites). The *matK* barcode region is also derived from the complete *A. thaliana* plastid genome sequence (gi 7,525,012:2056–3570) which is ~841 bp in length but at the center of the gene (starts at 205 bp and ends at 1046 bp, including primer sites).

Both *rbcL* and *matK* can be easily amplified from the plant genome and particularly, *matK* possess higher discriminatory power and generate adequate variation in even closely related plant species. It is also established elsewhere that among the actively evolving coding regions of plastid genome, *matK* is on the top and it is recognized as the most similar plant analogue to the *CO1* animal barcode. On the other hand, *matK* cannot be easily amplified by PCR using universal primers, particularly, it is the most difficult in non-angiosperms. Contrastingly, *rbcL* can be easily amplified from the plant species and their sequence can be easily aligned with the reference sequence. Though it produces relatively lesser discriminatory power, it can be used as a foundation to create a DNA barcode database for the given plant species.

Further, owing to the fact that these DNA barcoding regions are obtained from coding genes, in silico translation of the obtained sequences can be used to analyze the errors due to editing or assembly, orientation of sequences, and existence of pseudogenes. Such analysis offers the opportunity to assess the diversity in DNA barcodes among taxonomic groups that belongs to different geographical regions (Hollingsworth et al. 2011). At the same time, it should be noted that the discrimination power of *CO1* in animals is far better than *rbcL* + *matK* in plants, and it would be always suggested to employ additional barcoding genes, if greater resolution is required.

9.4.2 DNA Barcoding of Moringa Ecotypes Cultivated in Southern India

Keeping the above points in the workplan, this laboratory has generated DNA barcodes in Moringa using *Internal Transcribed Spacer* (*ITS;* which comprises both *ITS1* and *ITS2*), *rbcL,* and *matK* genes. Such combinations of DNA barcodes clearly provided sufficient discrimination among and even closely related Moringa ecotypes that are prevailing in Southern India. Resolution of *ITS* in discriminating few selected Moringa ecotypes cultivated in Southern part of India is provided in Fig. 9.2. All the *ITS* sequences obtained from these ecotypes were submitted for public use and it is available as GenBank IDs KT737744 to KT737801. When compared with *matK* and *rbcLa*, *ITS* has a greater number of single nucleotide polymorphism as evident in Fig. 9.3.

The phylogenetic tree constructed by *ITS* gene sequences classified all the investigated accessions into two groups (Fig. 9.4). As expected, group I was constituted by outgroup alone and rest of the investigated moringa accessions were grouped as group II. In this *ITS* phylogenetic tree topology, MO 53, MO 54, and MO 52 were grouped as clade I and the remaining accessions comprised as major clade II. The boot strap value was marked at the point of node divergence. The maximum divergence was noticed in node I among MO 53, MO 54, and MO 52. The highest conservation on the sequence of *ITS* was seen between MO 42 and MO 24 followed by MO 36 and MO 47 and MO 25 and MO 12 (Fig. 9.3).

On the other hand, it should be mentioned that several investigated moringa ecotypes cannot be distinguished using the *ITS* sequence. The low discrimination power of *ITS* and *matK* is already reported in *Allium* species and wild potatoes. However, it is suggested that use of more number of barcoding genes such as *rps16*, *trnH-psbA*, *trnL-F*, *trnD-T*, and *rpL32-trnL* can be utilized for better discrimination of very closely related moringa accessions such as those reported above.

Next-generation sequencing (NGS) kindled a transformation in DNA barcoding (especially

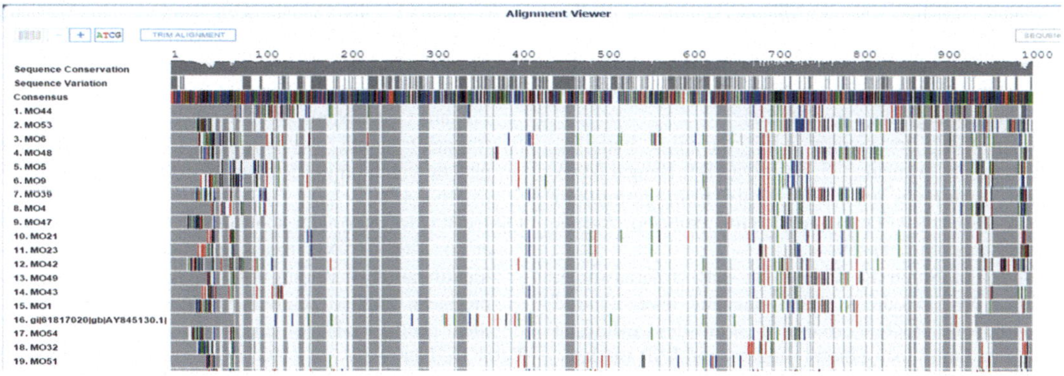

Fig. 9.3 Results of DNA barcoding of investigated moringa ecotypes using DNA subway by employing "MUSCLE" tool utilizing *ITS* gene sequence. Totally 60 genotypes (denoted with a prefix MO) were investigated in this study along with one outgroup and, barcodes of only 19 investigated barcodes were shown below along with the outgroup

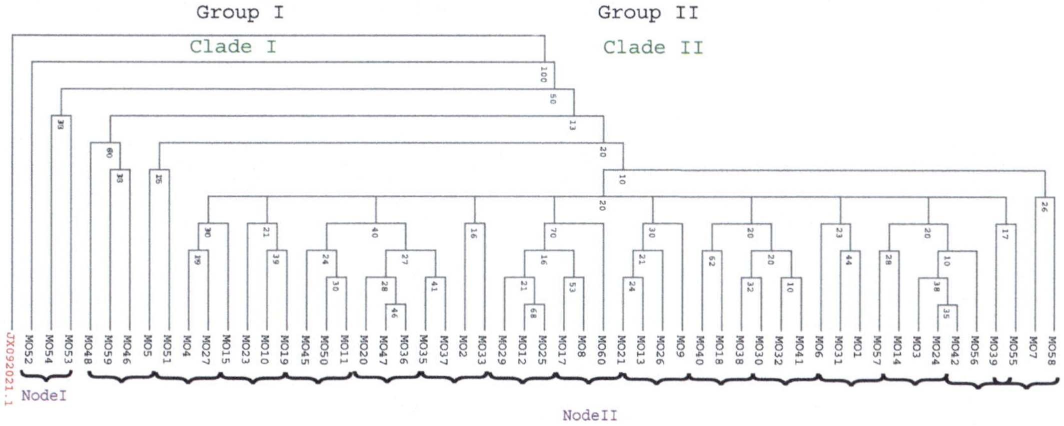

Fig. 9.4 Phylogenetic Tree constructed from the sequences of ITS amplified from the 60 moringa ecotypes grown in Southern India. Sequence of an outgroup (GenBank ID: JX092021.1) has also been included in the analysis to validate the results

metabarcoding) as it produces lengthy average sequence read length. Thus, huge amount of sequencing data can be generated from the several targeted regions of the investigated Moringa samples, and thus, it offers greater power to discriminate the samples at various genomic regions in addition to the above barcoding regions. Hence, the fundamental application of DNA barcodes, identification of unknown plant samples, gets more sophisticated and automated with the help of the NGS tools. This would be more desirable in herbal industry as raw materials used in herbal medicine are obtained directly from the markets as dried or powdered plant parts. Further, combining Moringa DNA barcoding with NGS, metabolomics, transcriptomics, and proteomics will facilitate species discovery, evolution, and conservation of useful variation for future.

References

Avila-Treviño JA, Muñoz-Alemán JM, Pérez-Molphe-Balch E, Rodríguez-Sahagún A, Morales-Domínguez JF (2017) *In vitro* propagation from bud and apex explants of Moringa oleifera and evaluation of the

genetic stability with RAMP marker. S Afr J Bot 108:149–156
Boopathi NM (2020) Marker assisted selection. In: Genetic mapping and marker assisted selection. Springer, Singapore, pp 481–504
da Silva AVC, dos Santos ARF, da Silva Lédo A, Feitosa RB, Almeida CS, da Silva GM, Rangel MSA (2012) Moringa genetic diversity from germplasm bank using RAPD markers. Trop Subtrop Agroecosystems 15(1):31–39
Ganesan SK, Singh R, Choudhury DR, Bharadwaj J, Gupta V, Singode A (2014) Genetic diversity and population structure study of drumstick (Moringa oleifera Lam.) using morphological and SSR markers. Ind Crops Prod 60:316–325
Harmon L (2010) Rare flowers and common herbal supplements get unmasked with plant DNA barcoding (reporting unpublished data of Damon Little and David Baker). Scientific American Observations. http://www.scientificamerican.com/blog/post.cfm?id=rare-flowers-and-common-herbalsupp-2010-04-18
Hassan FA, Ismail IA, Mazrou R, Hassan M (2020) Applicability of inter-simple sequence repeat (ISSR), start codon targeted (SCoT) markers and ITS2 gene sequencing for genetic diversity assessment in Moringa oleifera Lam. J Appl Res Med Aromat Plants 100256
Hassanein AMA (2018) Nutritional, chemical and molecular characterisation of Moringa oleifera Lam. and Moringa peregrina (Forssk.) Fiori genotypes. J Hortic Sci Biotech 93(5):500–509
Hassanein AMA, Al-Soqeer AA (2018) Morphological and genetic diversity of Moringa oleifera and Moringa peregrina genotypes. Hortic Environ Biotechnol 59 (2):251–261
Hollingsworth PM, Graham SW, Little DP (2011) Choosing and using a plant DNA barcode. PloS One 6(5):e19254
Kesanakurthi RP, Fazekas AJ, Burgess KS, Percy DM, Newmaster SG et al (2011) Spatial patterns of plant diversity below ground as revealed by DNA barcoding. Mol Ecol 20:1289–1302
Kumar P, Dolkar R, Manjunatha G, Pallavi HM (2017) Molecular fingerprinting and assessment of genetic variations among advanced breeding lines of Moringa oleifera L. by using seed protein, RAPD and Cytochrome P450 based markers. S Afr J Bot 111:60–67
Muluvi GM, Sprent JI, Soranzo N, Provan J, Odee D, Folkard G, McNicol JW, Powell W (1999) Amplified fragment length polymorphism (AFLP) analysis of genetic variation in Moringa oleifera Lam. Mol Ecol 8 (3):463–470
Ogden R, McGough HN, Cowan RS, Chua L, Groves M et al (2008) SNP based method for the genetic identification of Ramin (Gonystylus spp.) timber and products: applied research meeting CITES enforcement needs. Endanger Species Res 9:255–261
Patel JS, Narayana GV (1937) Chromosome numbers in some economic flowering plants. Curr Sci 5(9):479–479
Rajalakshmi R, Rajalakshmi S, Parida A (2017). Evaluation of the genetic diversity and population structure in drumstick (Moringa oleifera L.) using SSR markers. Current Sci 112(6):1250
Rajalakshmi R, Rajalakshmi S, Parida A (2019) Genetic diversity, population structure and correlation study in Moringa oleifera Lam. using ISSR and SRAP markers. Proc Natl Acad Sci India Sect B Biol Sci 89 (4):1361–1371
Ravi RD, Siril EA, Nair BR (2020) The efficiency of Cytochrome P450 gene-based markers in accessing genetic variability of drumstick (Moringa oleifera Lam.) accessions. Molecul Biol Rep 47 (4):2929–2939
Saini RK, Saad KR, Ravishankar GA, Giridhar P, Shetty NP (2013) Genetic diversity of commercially grown Moringa oleifera Lam. cultivars from India by RAPD, ISSR and cytochrome P 450-based markers. Plant Syst Evol 299(7):1205–1213
Tanksley SD (1983) Molecular markers in plant breeding. Plant Mol Biol Rep 1:3–8
Wu JC, Yang J, Gu ZJ, Zhang YP (2010) Isolation and characterization of twenty polymorphic microsatellite loci for *Moringa oleifera* (Moringaceae). HortScience 45(4):690–692
Zane L, Bargelloni L, Patarnello T (2002) Strategies for microsatellite isolation: a review. Mol Ecol 11:1–16

Genome Sequencing, Organellar Genomes and Comparative Genomics in Moringa

10

N. Manikanda Boopathi, R. Veera Ranjani, and M. Raveendran

Abstract

Consequent to the continuously increasing deposit of DNA sequence information from different DNA sequencing technologies, several novel avenues have opened to identify the genes and/or molecular markers linked to the agronomic and economic traits. However, the greatest tricky point in this genomics era is realizing the application of those huge DNA sequence data in the routine crop breeding programs. Thanks to next generation sequencing (NGS) tools and strategies, draft (reference) and resequenced Moringa genome data are now available for further mining, and it facilitates rapid identification of desirable candidate genes besides permitting to generate thousands of single nucleotide polymorphism (SNP) and other marker classes. Further, NGS can also be explored for unraveling the process of methylation and its implications in the genome and how such information can be used in Moringa improvement programs. Besides studies on organellar genomes and comparative genomics offer several other useful ways to improve Moringa for valuable nutritional and medicinal traits. This chapter provides a detailed description of the above genomics titles in Moringa.

10.1 Moringa Genome Sequence: Importance and Applications

The fundamental molecular unit of biological life is the genetic code. Acquiring the basic information on the genetic code such as the primary and secondary structures of DNA, how the genes are organized and regulated during gene expression is a key component in plant molecular biology. Such information would greatly benefit in designing a sound plant breeding program that will precisely speed up the process of developing an elite cultivar.

The discovery of genes in a crop plant (in which no primary information on DNA (such as genome sequence, chromosomal descriptions, and molecular markers) is available) is essentially dependent on uncharacterized genome sequence data and expressed sequence tags (EST). Initially, these data can be used to identify SNP and simple sequence repeats (SSR) markers which can be used to develop genetic maps, and consequently the genomic location of these markers can be identified without a reference genome sequence (see Chap. 9 for more details on molecular markers). Unfortunately, there are few such efforts in Moringa and still, there is a scarcity of identifying functional markers that

N. M. Boopathi (✉) · R. Veera Ranjani · M. Raveendran
Department of Plant Biotechnology, Center for Plant Molecular Biology & Biotechnlogy, Tamil Nadu Agricultural University, Coimbatore, India
e-mail: nmboopathi@tnau.ac.in

N. M. Boopathi et al. (eds.), *The Moringa Genome*, Compendium of Plant Genomes,
https://doi.org/10.1007/978-3-030-80956-0_10

can be used in designing efficient Moringa genetic improvement programs.

An alternative and powerful tool to identify the causal gene(s) of important agronomic traits is whole-genome sequencing of Moringa followed by bioinformatics analysis. Thanks to the rapid technological advances in DNA sequencing platforms (see below), draft or reference genome sequences of different crops (where genetic mapping has not been done) are now accessible, which provide opportunities to rapidly identify candidate genes through sophisticated bioinformatics tools besides discovering elite SNPs by comparing the reference genome with the investigated Moringa genome sequence.

Such effort of gene discovery in Moringa is considered relatively inexpensive and will offer an effective procedure not only to identify genes but also to identify gene promoters and other regulatory elements. It has been estimated that Moringa has 28 chromosomes (2n = 14) with a genome size in the range of 280–315 Mb (Tian et al. 2015; Chang et al. 2019). Applications of genome sequencing effort possess the following benefits:

1. Moringa is an important and affordable nutritional crop for the people living in the resource-limited environments. Precise genetic improvement of Moringa through genomics-enabled approaches will ensure and strongly improve the nutritional status and livelihoods of poor and neglected people living in resource-limited environments.
2. Due to enhanced awareness of healthy food products among global consumers, Moringa is now considered a major economic driver for some developing countries, especially for India, which annually produces several thousand tons of dried leaves and exports leaf powder and other Moringa value-added products worth several million dollars. Thus, increasing the productivity using the cutting-edge genome sequencing technology would boost the economic and social well-being of the poor farmers living in the rural areas.
3. Moringa is an outstanding model for the study of metabolic pathways that lead to synthesize a large array of nutritional and medicinal compounds (see Chaps. 3 and 4). Moringa leaves have been shown to be composed of sufficient amounts of proteins, vitamins, antioxidants, minerals, and other phytochemicals that meet the daily requirement of the human diet in an affordable way. Translational genomics of such metabolites may lead to the improvement of not only Moringa but in other diverse biomass crops.
4. *Moringa* genus comprises 13 species, and decoding Moringa genomes will shed light on the functional and agronomic significance of different genomes *Moringa* genus (in terms of genome size and sequence variation), besides unraveling the hidden perspectives of evolution and acclimatization to diverse climatic conditions.
5. The Moringa genomic resources allow conducting imperative functional and evolutionary genomics studies that may facilitate genome-enabled targeted trait (demanded by stakeholders) improvement in Moringa. Plant genomics studies would be a valuable tool to understand important traits such as flowering and fruit formation in Moringa besides their role in increased biomass production.

10.2 Approaches Used for Whole Genome Sequencing in Crops Including Moringa

Challenges in plant whole genome sequencing include covering the repeat-rich regions of the genome, variations in genome size due to ploidy level, heterozygosity due to cross-pollination, historical and recent mutations due to transposable elements, etc. Though polyploidy and heterozygosity limits the efficient assembly of whole-genome, such limitations have opened up new avenues in development of novel approaches in whole-genome sequencing such as haplotype phasing, structural variant analysis, and de novo pan-genomics. Further, the strength enabled by the de novo transcriptome assemblies facilitates a complementary approach that avoids the complexity arising during whole-genome assembly and explores the gene pool of polyploid plant species.

Haemophilus influenzae was the first de novo sequenced genome of a free-living organism, and it was done in 1995 using Sanger sequencing technology and demonstrated the utility of whole-genome shotgun sequencing (WGS) in genome sequencing. WGS involves randomly fragmenting the genome and cloning it into plasmid vectors, paired-end sequencing (see below), and assembling the sequence reads using powerful algorithms. Demonstration of such a WGS approach can produce three kinds of raw reads of sequences. A general scheme of producing such read is detailed below:

(a) Single-end reads: Initially each of the randomly fragmented whole genomic DNA is ligated with an adaptor and an adaptor-specific primer is used to sequence the fragment. Thus the sequence has resulted from a single end, i.e., the adapter end. However, for Sanger sequencing, the fragmented DNAs are cloned into a plasmid vector and the primers are designed in such a way that they flank the cloning site.
(b) Paired-end reads: Similar to the above procedure the adapter is ligated to the sheared whole-genome DNA fragments, but the sequencing is done in two different reactions using different primers that flank both ends of the target DNA. However, the adapter sequences at each end of the fragment are having different priming sites. Thus, sequence data on both of the complementary strands are obtained and offer a guarantee during the alignment.
(c) Mate pair reads: After creating the sheared whole genomic DNA fragment library (with the fragment size of longer than 1 kb), each fragment is circularized around a single adapter and thus the fragment ends are ligated to the adapter ends. Those circularized molecules are digested with type II-S endonuclease or modified by nick translation to release a single linear fragment which now has the original DNA fragment on both ends but an adapter at the center. All the other DNA traces are removed by capturing only the central adapter as it is biotinylated and can be captured using streptavidin magnetic beads. Adaptors are ligated to these selected fragments, and they are sequenced (either using single-read mode (454 Roche platform) or paired-end mode (Illumina or SoLiD™ platforms); see below for details on different sequencing platforms). As the separation distance is longer, the resulting reads are aligned as a pair to the genome of interest which provides relatively larger coverage on the genome.

Accordingly, the WGS procedure in the early 1990's mainly involved fragmenting the genome into small DNA segments and inserting them independently into vectors such as cosmids, bacterial artificial chromosomes (BAC), and yeast artificial chromosomes (YAC). Each clone is then sequenced individually, and the sequences of each clone were assembled to form consensus sequences. The consensus sequences are then arranged to form pseudomolecules which ultimately generate contigs. As a final point, all the associated contigs are consecutively put together into scaffolds and individual chromosomes.

Thus, during the last couple of decades, the strategies employed to WGS changed swiftly and sequencing platforms and bioinformatics tools used to assemble the genome have become sophisticated extensively since the first plant genome sequence, *Arabidopsis thaliana,* published in 2000. It was sequenced by employing Sanger-based BAC—by—BAC strategy with a consortium of global scientists over years. Even though this strategy offered high-quality chromosome-scale reference sequences, such efforts were expensive and labor-intensive.

On the other hand, it is now possible to obtain a high-quality and precise *Arabidopsis* reference genome within a week by involving few scientists with a minimum tiling path of bioinformatics pipelines.

Bioinformatics tools used for the sequencing also changing rapidly: an approach that involved shotgun sequencing combined with an overlap layout consensus (OLC) strategy and CELERA assembler (which was formerly developed for the assembly of the human genome) resulted in high-quality assemblies at a fraction of the cost and time, and it has been used in earlier studies in papaya, soybean, and poplar genome assembly.

However, later when the second generation sequencing technologies such as 454 and Illumina (see below) were developed, it motivated the designing of the De Bruijn graph (DBG) assembly methods (see Chap. 12) that can deal with shorter reads sequenced at greater depth with simple effort and affordable cost. Despite the fact that the second generation sequencing method produces lower quality assemblies, such efforts have discovered several genome-enabled discoveries in numerous economically imperative crops.

After 2015, single-molecule sequencing strategies once again revolutionized the methods of plant genome assembly and enabled to assembly close to complete chromosomes. *Oropetium thomaeum,* a desiccation-tolerant grass, was the first plant genome that was sequenced by Pacific Biosciences (PacBio) single molecule real time (SMRT) sequencing method. This sequencing strategy led to the development of Falcon, a novel assembler to gather long error-prone reads. There is also an additional long read assembly software, CANU, which was actually an update to the CELERA assembler.

Besides, there are few more technologies that have the promise of completing the chromosome level assembly without using the time-consuming and expensive construction of physical map through BAC clones or genetic maps with molecular markers. These include high throughput chromatin conformation capture (Hi-C) and optical maps and nanopore sequencer introduced by Oxford Nanopore Technologies (ONT). In some cases, ONT surpassed PacBio in producing read lengths (exceeding mega bases; Mb) and it has shown its utility in releasing complete reference genomes of *Arabidopsis*, Tomato, Sorghum, Banana, and Brassica.

Thus, the approaches used for plant genome sequencing are ever changing as the technology develops. The Moringa sequencing efforts have started in the middle of the plant genome sequencing era, and hence the initial approach that was employed to sequence the Moringa genome was focused on shotgun sequencing as detailed above.

10.2.1 Sequencing Platforms

10.2.1.1 Evolution of Genome Sequencing Concepts

Since the dawn of Sanger's dideoxynucleotide sequencing in the 1970s, novel innovating sequencing technologies are continuously growing to unravel the biological unanswered questions using the genomic data.

Initially, RNA sequencing was the focus of Dr. Frederick Sanger's laboratory at the Medical Research Council in Cambridge, United Kingdom, and then he focused on a DNA sequencing method. He employed primer and DNA polymerase to synthesize new strand based on the DNA template information, but with usual dNTPs, he also added ddNTPs (3' hydroxyl group of dNTPs are replaced with a hydrogen atom; such replacement did not allow nucleotide chain elongation as the hydrogen ion is functionally incapable to react with the next nucleotide and synthesizing new strand is stopped).

Thus, the addition of four native dNTPs with one of the four of their ddNTP analogs resulted in successive nucleotide-specific terminated DNA fragments for each of ^{32}P labeled molecules, *viz*., dATP, dTTP, dGTP, and dCTP. The nucleotide-specific terminated DNA fragments are then resolved based on their size (at single-nucleotide resolution) on thin slab polyacrylamide gels by running each reaction of the A, C, G, and T in successive neighboring lanes to visualize the position of each nucleotide on the fragment after exposing the dried gel into the X-ray film. Because of the X-ray documentation, it is now simple to read the gel image from bottom to top (i.e., from the shortest to longest amplified DNA) and hence the sequence of the DNA fragment can be obtained (Fig. 10.1).

Though it is an elaborate and labor-intensive process, such effort has helped in further DNA sequence downstream applications such as

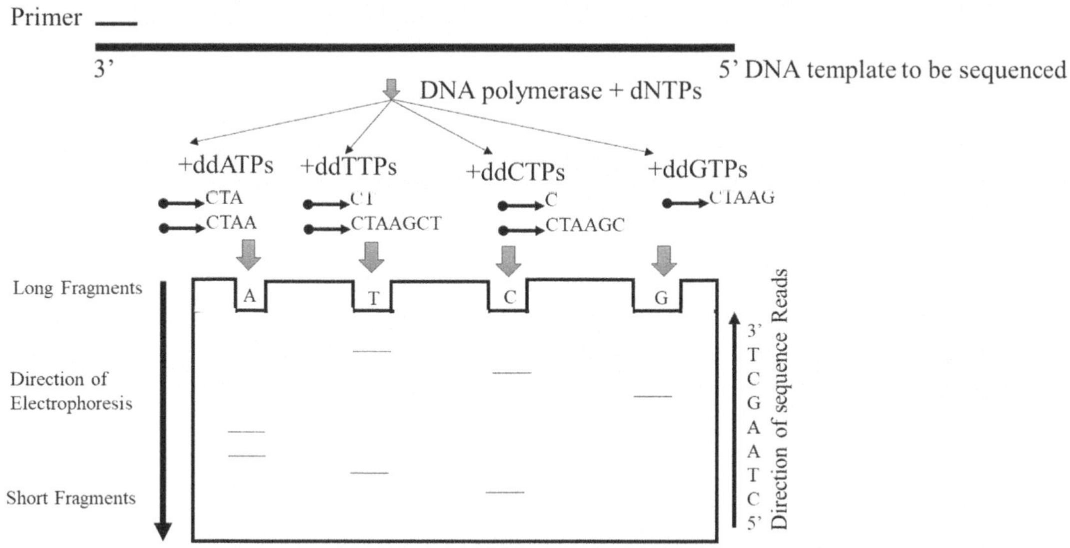

Fig. 10.1 Schematic illustration of Sanger dideoxynucleotide sequencing strategy, a first of its kind DNA sequencing method

gene/genome assembly and translation to amino acid sequence during the late 1970s to 1990s.

Alternatively, Maxam–Gilbert or the chemical method of DNA sequencing was introduced in 1977 by Allan Maxam and Walter Gilbert. It is mainly based on the chemical modification of each nucleotide in the target DNA followed by cleaving that modified DNA exactly at the site that is adjacent to the modified nucleotides. However, this method involves series of chemical reactions and the success largely depends on the completion of such reactions. As further advances in sequencing enzymology and processes were introduced in the late 1980s, the chemical sequencing method lasts its widespread applicability.

Introduction of fluorescently labeled dNTPs (different colored fluor for each nucleotide) instead of radiolabeled nucleotides by Leroy Hood's laboratory at the California Institute of Technology and marketing of a fluorescent DNA sequencing instrument by Applied Biosystems, Inc. (ABI) in 1986 has revolutionized scalability and affordability of the DNA sequencing technology. Automated reading of the DNA sequence by scanning the gel plates with laser beam leads to excitation of different wavelengths for each labeled fluorescent dNTPs that were size separated during the electrophoresis. Thus the process of radiolabeling the ddNTPs, meticulous gel drying, exposing and developing hazardous X-ray film, and manually reading the autoradiograph was eliminated and enabled DNA sequencing as a simple and affordable procedure.

However, the painstaking process of slab gel casting and electrophoresis of the DNA fragments limits the speed of the DNA sequencing process. This hurdle was overcome by the introduction of capillary electrophoresis in 1999 first by MegaBACE™ sequencer from Molecular Dynamics and then the ABI PRISM® 3700. DNA fragments are resolved on a polymeric separation matrix that is directly injected into capillaries which provides single-nucleotide resolution.

Further advancements in this process (such as loading the samples directly from the microtiter plate to the capillaries using electrical current pulses (a process known as an electrokinetic injection), accelerating run times by rapid heat degeneracy of the capillaries, replacement of polymer matrix with new matrix pumping once the separation and detection process is over, fixing the capillaries in their positions in order to facilitate much faster and more accurate data extraction and base-calling) have avoided several time and effort consuming steps and simplified DNA sequencing method. In fact, the majority of reference genomes of the plant models and humans were produced using ABI PRISM 3700 and its updated version ABI 3730 possessed the above advancements.

A ground-breaking change was occurred in DNA sequencing during 2005 due to the introduction of next generation sequencing (NGS) instruments and strategies. Sanger sequencing employed construction of large clones followed by creating smaller subclones, sequencing the subclones, and assembling those sequences to obtain the sequences of larger clones, the larger overlapped clones are used to prepare contigs and finally restructure the chromosomes (which may be called as a top-down approach of sequencing). On the other hand, NGS not at all necessitates the cloning step itself.

Instead, as a preparative NGS, the target genomic DNA is sheared to create a large library of fragments which are covalently attached by DNA ligase to adapters (synthetic DNAs specific to each NGS platform, which act as universal primers in the consequent amplification of each fragment of the library) added to each fragment end.

NGS also differ from the Sanger sequencing in another way: it does not need to do the sequencing reactions in the minimum of four microtiter plate wells; instead, amplification of the DNA fragments are carried out in situ on a solid surface (using a bead or flat glass microfluidic channel which is covalently linked with complementary adapter sequences that are used for the library preparation). Such amplification process is actually a single focused digital process and each bead may contain a cluster of amplified DNAs, which are originated from a single fragment of the library. This amplification is digital in nature as each amplified fragment yields a single focus (a bead- or surface-borne cluster of amplified DNAs, all of which originated from a single fragment). Thus, necessary quanta of signals are produced during amplification which leads to generating corresponding sequencing data specific to each library fragment.

In contrast to the Sanger sequencing where the nested fragments are arranged to find the actual sequence of a single fragment, NGS is having massively parallel sequencing effort and due to that the scale and throughput of sequencing are increased several folds higher than the Sanger sequencing. The steps involved in massively parallel sequencing are

1. Addition of nucleotide
2. Detection of incorporated nucleotides on each target fragment and
3. Washing of unused fluorescent labels or blocking groups.

Actually, sequencing and detection in NGS are a simultaneous process and that too they occur in a massive manner (in the given point of time hundreds of thousands to billions of reactions can be performed by a single NGS instrument, which leads to generating massive data sets (in several hundred mega base pairs; Mb).

The final read length is also distinguishable between Sanger sequencing data and NGS. The highest read length can be obtained in Sanger sequencing as the read length is largely dependent on factors that determine gel electrophoresis (such as the percentage of polyacrylamide, the electrophoresis conditions, the time of separation, and the length and thickness of the gel). On the other hand, shorter read lengths are obtained in NGS due to the fact that read length is largely dependent on a good signal-to-noise ratio which is greatly different among the NGS technologies. Thus, even though NGS generates a large amount of sequence data, the shorter read length introduces difficulties during the assembly of contiguous sequences. Consequently, it leads to large gaps in the polyploid plant genome, as it has repetitive elements and gene families.

Fortunately, Moringa does not have such limitations in generating NGS genomic data as it has a smaller genome (the published reports stated that it has a maximum size of ~315 Mb (Tian et al. 2015)), which is similar but smaller than rice (*Oryza sativa*) genome. Moreover, as premium quality reference genome sequence data and precise algorithms (see Chap. 12) have already been established in model organisms including Moringa (see below), now assembly of shorter read lengths produced by NGS can easily be realized. Similarly, advancements in NGS tools that focused on fine-tuning of signal-to-noise ratio as well as paired-end sequencing (sequencing both ends of each fragment) have resulted in increased read length from NGS.

During de novo sequencing (sequencing genome for the first time) and assembling the complex regions of a genome (regions rich in repeats), the combination of mate-pair and paired-end reads (see Sect. 10.2) are employed to enhance the genome coverage as the mate-pair reads offer longer-range order and orientation (with a separation distance of ~20 kb) and the paired ends offer potential to assemble several overlapped contigs to scaffold.

10.2.1.2 Development of Different Commercial NGS Platforms

Illumina

The first instrument system of Illumina® Inc (which used reversible dye terminators for high-throughput sequencing of DNA fragments) was originally designed by Solexa in 2007 and was later purchased by Illumina® Inc. Sequencing proceeds with fragmenting the high–molecular weight DNA, trimming them with enzymes followed by adenylation of the ends that facilitate ligation of proprietary adapters.

There are eight microfluidic channels in the Illumina microfluidic conduit, which is actually a flow cell, and each channel is having a complementary sequence that can efficiently bind to the target DNA which has already ligated with adapter sequences. Thus, using a precise setup of library concentration, the target DNA will be amplified in situ on each channel as clusters and they are sequenced. As the blocking nucleotide (3'H at the ribose sugar) exists in the amplification medium, amplification stops in each nucleotide in different clusters which will be identified by the reversible fluorescent label. The overall illumina sequencing comprises the following steps:

1. Addition of successive nucleotides by polymerase
2. Removal of unincorporated nucleotides
3. Imaging the fluorescent signal from both inner surfaces of the flow cell and identify each cluster nucleotide
4. Cleaving the fluorescent groups using proprietary chemicals and
5. Deblocking the 3'OH.

The above steps are repeated to identify up to 150 nucleotide addition reactions, and then the second read is subjected to the same steps. A similar kind of read is obtained from the opposite end of each fragment, but before that, the system eliminates the already synthesized strand by denaturation and using different cleavage reagents. These entire steps are automated in the flow cell and produce quality forward (first) and reverse (second) reads of the target DNA.

It is also informative to note that 0.5% substitution errors may occur during the sequencing process (finding a wrong nucleotide in the read, i.e., one error in 200 bases).

From 25-bp single-end reads produced by the original Solexa instrument, now the read lengths have increased to 150-bp paired-end reads (Illumina HiSeq 4000). This read length development has led to increased throughput in a day (from 1 Gb for the Solexa 1G to 600 Gb for the HiSeq 2000).

Ion Torrent

During DNA amplification, the addition of nucleotide releases hydrogen ions, which lead to quantifiable changes in pH. This pH change can be detected using a unique silicon detector available in the Ion Torrent instrument system and based on this, the DNA sequence is enumerated. This principle was commercialized in 2010 by Ion Torrent, which was subsequently acquired by Life Technologies™ Corp. The general procedure of this strategy includes fragmenting the target DNA, polishing the ends with enzymes that facilitate the ligation of adapters.

A unique strategy, called emulsion PCR, is used to amplify such library fragments. This strategy quantifies the library fragments, dilutes them, and mixes them in equimolar quantities with small beads (which are covalently linked with adapter complementary sequences), PCR reactants, and DNA polymerase. Vigorous shaking of this mixture leads to the development of an emulsion (in which beads and DNA are encapsulated in oil micelles (at 1:1 ratio on an average) that facilitate thousands of concurrent PCR-based amplification in one vessel.

Subsequently, oil is removed (which is called emulsion breaking) and enriched with PCR amplified beads by removing the beads with unamplified DNA. After adding the sequencing primer, the beads are placed on the silicon chip (referred to as Ion Chip) which can detect the pH changes, as the reaction progresses.

Though the Ion Torrent sequencer does not have noises as it uses proprietary nucleotides for the sequencing reaction, noise may be accumulated owing to the phasing (i.e., at each step, all the fragments are not extended by nucleotide incorporation). Thus, it may lead to a 1% error rate (i.e., 1 in 100 bases). As the single read length of this instrument has increased from 100 to 200 bp, the throughput has increased from 10 Mb per run to 1 Gb per run.

In general, the Ion Torrent sequencer is recommended to the laboratories with limited resources but needs to use NGS, as it demands little reaction volumes and less investment in instrument procurement (due to mass production of the Ion Chip using standard semiconductor techniques).

PacBio

Pacific Biosciences Inc. (PacBio) has introduced single-molecule sequencing of DNA using zero-mode waveguide (ZMW) in 2010, which has perfect detection of even a small amount of fluorescence. ZMW is a light-focusing structure, and thousands of such ZMWs are spotted on a silicon wafer surface.

It also uses specific DNA polymerase that has the ability to sequence a single molecule with a reduced rate of incorporation of nucleotides that are labeled with fluorescent molecules. This leads to exceptionally sensitive detection of fluorescence from a single molecule of DNA in real time (i.e., whenever a nucleotide is incorporated by the DNA polymerase).

The workflow in PacBio includes fragmenting the genomic DNA into small DNA library fragments. The fragments are polished in such a way that they can be ligated with hairpin adapters. After denaturation, the hairpin adapter helps to form a circle comprising adapter–Watson strand–adapter–Crick strands. Later, they are deposited on ZMWs (SMRT Cell) after attaching DNA polymerase to each fragment. Primer, which is complementary to the adapter, is annealed to the library molecules with appropriate concentration. DNA polymerase attaches to the bottom of each ZMV where it has excitation and detection optics that has the ability to detect every fluorescently labeled nucleotide in real time as the optics repetitively survey the active site of DNA polymerase in every 150,000 ZMWs on the SMRT Cell. As stated above, the DNA polymerase used here are having a slow rate of incorporation of the nucleotide, it provides sufficient time to detect the fluorescent signal from each nucleotide.

As the pyrophosphate is labeled with the fluorescent dye, it diffuses away once it is released from added nucleotide and hence it helps to unambiguously monitor each nucleotide addition by each ZMW. Thus, large data is collected as clusters which are further used to get the sequencing reads.

As read length largely depends on the size of the library fragments and the time of data collection, PacBio can lead to producing up to 10,000 nucleotides at maximum. However, it has an overall higher error rate in single-molecule sequencing (owing to i) omission of fluorescent labeling may occur during labeling and subsequent purification steps and ii) though the nucleotides are detected, it may not be incorporated due to poor polymerization event or *vice versa*). These and other error factors have resulted in an error rate of approximately 15% (15 bases per 100 bp).

On the other hand, PacBio has helped to analyze the methylated DNA, which is considered an important regulatory mechanism in the genome. Besides, it has also been shown to be useful in measuring the ribosomal translocation on a messenger RNA and structures of each protein receptor on the plasma membranes at single-molecule resolution.

From the above, it is clear that due to problems associated with library preparations and errors generated by DNA polymerase, and there is a need to develop a platform that can overcome the aforesaid challenges and increase the real read length.

Oxford Nanopore Technologies (ONT)

MinION commercialized by Oxford Nanopore Technologies (ONT) is the first platform that used nanopores for DNA sequencing. It has the ability to differentiate the individual nucleotides by measuring the change in electrical conductivity when DNA molecules surpass through the nanopore.

The most important points in the ONT sequencing platform are (i) it does not depend on sequencing by synthesis and hence there is no need for PCR amplification and its associated errors (ii) it results in accurate long reads in real-time that can be aligned to the reference genomes, and (iii) as of date, this is the most portable DNA sequencer available in the market with lowest investment cost. However, MinION still needs a library preparation step, before proceeding to sequence, which is required to be optimized for each sample, as outlined below.

Hairpin structures are ligated to the end of double-stranded library fragments and then denatured to produce one length of single-stranded DNA comprising forward strand followed by the hairpin sequence then the reverse strand. After processing, the MinION produces three types of reads for each library fragment: Template, Complement, and Two Direction. All these data were used to provide a consensus sequence of the Template and Complement reads using an ONT base-calling software, and this sequence is called a Two Direction Read.

It should be noted that all the three read types are not always produced by the ONT and only few nanopores produce such combined reads. Few of the nanopores result in Template read alone and others in Template and Complement.

However, both the third-generation sequencing (TGS) technologies (PacBio and ONT) can generate long-reads as a single molecule and provide information on the native DNA as much as possible which facilitates filling the gaps that occurred during short-read assemblies by NGS and more importantly the loci that are methylated.

By comparing the two TGS technologies, it can be concluded that establishing a MinION platform in the small-scale laboratory seems to be cost-effective and more productive as it involves (i) small portable devices for monitoring the progress of the sequencing reaction in real-time and detecting nucleotide modifications and (ii) does not involve huge instrumentation facility and complicated library procedures.

On the other hand, MinION produces a high error rate (for example, Laver et al. (2015) reported an error rate of 38.2%, after base calling). This suggests that involving both NGS and MinION sequence reads will enhance the contiguity of de novo assembly. The other limitations associated with TGS are listed out in Xiao and Zhou (2020). TGS required fresh material or intact cells and preparation of ultra-long high molecular weight DNA for DNA library construction. When compared with NGS, the cost of TGS is still higher ($65–$200 per Gb in the PacBio and $22–$90 per Gb in the ONT). TGS also needs to be improved and challenging bioinformatics pipelines as it needs specific assembly algorithms and database systems for interpreting complicated sequence variations.

Thus, with the available funding opportunity and infrastructure facility, it is important to decide which sequencing platform (NGS vs. TGS) is to be selected, before starting a new plant genome sequencing project. As each platform has its own read length limit and we have a broader spectrum of read length with different platforms (Sanger sequencing ($\sim$1 kb sequence reads) and Roche 454 sequencing (up to 800 bp), Illumina HiSeq (at present typically 150 bp) and SOLiD (typically 50 bp), Pacific Biosciences (up to 5 kb), IonTorrent ($\sim$500 bp) and Illumina Moleculo (up to 10 kb)), it is imperative to have a meticulous planning that has the project objective in mind. As of date, it is recommended to employ both NGS and TGS platforms as the limitations of each platform are compensated although such a combination costs high and requires large skill sets.

10.2.2 Reference Genome Sequences

Recent interest in the development of reference genome sequences of *Moringa oleifera* was mainly due to its potential as a model tree crop for understanding basic biological processes and metabolism of health-promoting phytocompounds such as vitamins, minerals, antioxidants, and other medicinally important compounds. As Moringa has relatively short cycling times among the tree crops (six months, seed to seed), small tree size (at maturity, the tree can be maintained at the height of 2 m) and can be efficiently propagated using effortless vegetative propagation techniques such as cuttings and air-layering, it has been recognized as an excellent model tree crop for genome sequencing. Further, due to its diploid nature, Moringa facilitates trouble-free dissection of the genetic basis of important trits including oil synthesizing and biofuel crops.

To date, no genome sequence of Moringa wild species has been found and hence the domestication of Moringa has been poorly established with molecular data. Hence, rese quencing of Moringa accessions found in diverse origins would demonstrate how the Moringa plants from North-west India proliferated worldwide through the extensive introduction.

In addition, in the *Moringa* genus, several natural populations and mutagenized populations have been generated and conserved. Hence, the availability of a high-quality, well-annotated reference genome sequence in Moringa can be extended to develop super pangenomics toolkit with enriched physical maps that contain information on important genes involved in biotic and abiotic stress resistance as each *Moringa* spp., has evolved in extremely different environmental conditions.

It has been shown that there exists significant genetic diversity among the cultivated Moringa accessions and their wild accessions (see Chap. 7). Comparatively higher genetic diversity among the Moringa accessions may be either due to the continuous introgression of desirable traits from landraces into the compatible cultivars or due to the adaptation of Moringa cultivars to the new environments when they introduced. Although there is little genetic bottleneck during the domestication and artificial selection processes, genetic diversity was lost to some extent in elite Moringa cultivars. Therefore far-reaching sampling of wild and ancient Moringa accessions would be useful to unravel a comprehensive picture of Moringa super pangenome.

Such resources can also be efficiently compared with the genomes of other related trees as well as vegetable crops to reveal the common as well as matchless properties that helped for better adaptations to lead their life on earth. Such comprehensive comparative studies using Moringa as model genome can also exhibit the value of employing an intimately related diploid genome sequence during assembly of the complex polyploid genome and unraveling the molecular basis of cellular activities and other growth and development features related to any plant biological process.

Moringa was documented to originate from North-west India (see Chap. 1) and has spread to almost all tropical and sub-tropical climates including some parts of temperate regions. Since its introduction, in order to adapt to divergent habitations, Moringa has undergone a various natural selection and artificial varietal improvement procedures. Particularly, recent Moringa breeding programs mainly focused on genetic improvement of biotic and abiotic stress resistance and pod quality traits (good flavor, taste, and health benefits).

Such continuous procedures have resulted in the elevated concentration of secondary metabolites in the Moringa leaf or pod in the Moringa cultivars compared with their wild relatives. Therefore, understanding the genomic basis of Moringa quality phytocompounds has captivated persistent awareness among phytomedical and seed industries as this study would shed light on molecular mechanisms underlying Moringa plant secondary metabolism, stress resilience, and quality health benefits.

Table 10.1 Details on draft Moringa reference genome sequences available in the public domain and other studies

Particulars	Tian et al. (2015)	Chang et al. (2019)	https://www.ncbi.nlmnih.gov/bioproject/?term=PRJNA268707	Boopathi et al., (under preparation)
Name of Moringa germplasm accesions	Unnamed drumstick tree	Unnamed drumstick tree	Cultivar KDM-01 (Bhagya)	Cultivar PKM1
Genome size (Mbp)	315.00	216.76	253.9	215.33
N50 (bp)[a]	611	1295	2042	17,636
Number of predicted genes	19,465	18,451	18,203	41,801
Completeness evaluation of genome assembly using BUSCO database[b] (%)	–	88.8	–	78.82

[a]N50 value is defined as the length of the shortest contig in the set of largest contigs that together constitute at least half of the total assembly size. In general, a high N50 value signifies a high-quality draft assembly
[b]Genome completeness analysis was performed using Benchmarking Universal Single Copy Orthologs (BUSCO) database version 3.0.1

Though the recent surge in functional and structural genomic tools has provided huge information on the wide range of models on crop genome evolution and breakthrough in the identification of imperative genes involved in the expression of agronomic traits in annual crops, our knowledge of Moringa genomics and population genetics is poor and requires immediate attention due to its importance in food and nutritional security.

Further, the development of the Moringa reference genome will offer potential ways to study the evolutionary routes of transposable elements (TEs) which have several applications in plant genetic improvement. TEs are involved in the expansion of plant genomes and have strong regulatory control over expression as well as the structural organization of genes (Xia et al. 2020).

Although the draft genome sequences of two Moringa accessions (Tian et al. 2015 and Chang et al. 2019) provided the primary information on genome structure, evolution, and the genetic basis of medicinal and other health benefits of Moringa, further studies are required to provide information on completeness and quality of the reference genome assembly (Table 10.1). This greatly hampers the studies that focused on genome architecture and gene makeup of Moringa, comparative genomics, structural variation, and population genetic studies.

Therefore, understanding the genomic variants with improved genome assembly and resequencing diverse Moringa germplasm accessions would provide potential procedures in genetic improvement of Moringa for several economically and medicinally important traits.

10.2.3 Whole and Targeted Genome Sequencing and Assembly

Though recent advances (such as genotyping-by-sequencing, RAD-Seq, reduced representation sequencing, amplicon sequencing, and transcriptome sequencing) in the development of genetic markers and genome-wide analysis has made rapid progress in understanding genetic variation and identification of major candidate genes besides unraveling the basis of inbreeding depression at various levels such as epistasis, dominance versus overdominance, many versus few loci and providing insights on the effect of genetic variation on patterns of gene expression, resilience to environmental change and taxonomic delineation, a complete and well-annotated genome sequence offers several benefits in the crop improvement program.

Whole-genome sequencing data annotated with all inclusive genomic variation such as single nucleotide polymorphisms (SNPs), sequence repeats, insertion–deletion structural rearrangements, and copy number variation would be useful to unravel most useful genetic information (such as desirable haplotypes, comprehensive estimates of linkage disequilibrium, timing of admixture events, and advantageous signatures of functional regions for selection) which are useful to design an informative molecular breeding strategy.

Further, complete or draft reference sequences are also essential for characterizing isoform-specific and allele-specific gene expression using RNA-seq, understanding epigenetic modifications (such as methylation), and DNA–protein interactions.

However, the prior decision on the purpose of the whole genome sequencing is necessary to proceed further as it involves enormous resources, expertise, and time to assemble complex plant genomes (the key challenge is the repetitive sequences, polyploidy, and heterozygosity). Otherwise, the availability of genome sequence would be simply a resource and as such, it would provide little information. To tap the complete potential of whole-genome sequencing, the laboratory must have adequate funds and access to computational resources that are developed specifically for the given plant genome sequence. Instead of obtaining poorly covered whole-genome sequencing data (which is considered a waste of funding, effort, and time), it is suggested to employ other targeted genomics approaches that provide useful genomic information such as genotyping-by-sequencing and transcriptome sequencing.

Another point to be considered in plant whole-genome sequence data assembly is that the core genes (that are identified using the conservation strategy) are highly polymorphic and may have many paralogs. This situation complicates the genome assembly and such situations are also arise when there are rapidly evolving genes or members of large gene families. Consequently, these groups of genes are inadequately found in the final assembly and annotated gene set. Even though the aforementioned difficulty can be avoided by employing TGS (see Sect. 10.2.1.2) which results in very long reads, the challenge will exist if those regions are not correctly represented by conservation biology during the assembly.

To this end, employing targeted sequencing of BAC clones and careful manual annotation is required to comprise such genomic regions in the whole genome assembly. Therefore, in order to capture the physical position(s) of the already identified candidate genes, it is suggested to focus only on those targeted regions, instead of going for whole-genome sequencing.

Though the draft whole-genome sequence represents the complete nucleotide bases of all the chromosomes in the given species of interest, there will be different complications when we employ such sequences:

1. It represents only one true sequence for the investigated species and explains only one kind of variation in that taxonomic classification. On the other hand, the individual has a genetic variation (even if it belongs to diploid), at heterozygous positions, insertion/deletions (InDels), copy number variation, or small-scale rearrangements in each chromosome sets. Further, owing to somatic mutations, cells of the same individual may have variations in their genomic content.
2. Generally, though a single individual is sequenced, in some cases the draft genome represents consensus sequences of a number of pooled samples (example: International Human Genome Sequencing Consortium 2004). In contrast, the whole-genome assembly of the diploid and polyploid organisms (such as plants) represents the consensus sequence of several chromosome sets. Such assembly might have not captured the actual haplotypic variation using the short-read-based method, which is a common strategy used in the published plant genomes.
3. Typically it is impractical to sequence and assemble all nucleotides in the genome especially the heterochromatic regions found in centromeres and telomeres and other highly repetitive regions.
4. Each genome assembly may have doubts in its authenticity at least in certain parts of the

genome attributing to the errors that are stemmed from either during the sequencing process or during the assembly pipeline. Further, the introduction of ambiguous bases or 'N's in the sequence due to several limiting factors of NGS and TGS technology also hinders the further applications of these genome sequences.

5. Indeed, the whole genome sequence should be considered as a working hypothesis as it has resulted from a series of assembly errors. For example, the scaffolds that are formed by joining overlapping contigs are assigned to linkage groups or placed on chromosomes. For this purpose, genetic maps are frequently employed which are constructed from pedigree data or controlled crosses. However, the construction of informative genetic maps requires extensive genotyping of a suitable mapping population that has a significant amount of recombination events (Boopathi 2020).
6. Alternatively, gene(s) or regulatory element (s) synteny and their order obtained from similar plant species are used to place and orient the scaffolds onto chromosomes. Nevertheless, such effort may also lead to having erroneous conclusions due to the fact that chromosomal rearrangements are very common among closely related species and/or even within the plant species. Besides transferring the errors in those species will also be transferred to the given plant genome sequence.

Owing to the aforesaid limitations, it is realistic to define the purpose of the plant genome sequencing before starting the whole genome sequencing project.

10.2.4 Genome Assembly

Assembly of contigs using the overlapping reads is the key for further downstream applications, and it is found simple in the case of earlier methods such as Sanger sequencing which results in relatively long-read data of every distinctive BAC clone even with a limited sequencing depth.

In contrast, NGS produces short reads and hence it requires relatively higher read coverage; preferably more than 10× sequencing depth is used for diploids, and the sequencing depth varies with the genome size and ploidy level. Failing to do so, the resulting short reads and excessively little data instead of the prescribed sequencing depth will lead to fragmented assembly, which may cause issues during annotation and variant calling (see below).

Contig assembly starts with a high amount of paired-end short-read data and consequently, these contigs are assembled into scaffolds. For scaffolds preparations, the development of target DNA libraries with long-insert sizes (3–40 kb) is suggested to get sufficient overlapping read lengths. Factors that decide the amount of sequencing data required for each library type and insert size vitally depends on:

1. Genome size and its repeat content
2. Amount of heterozygosity present in the genome
3. Final quality required for genome assembly which is decided by the objective of the WGS program.

Thus, WGS and assembly process greatly differ among the sequencing projects and organisms of interest, and each assembly pipeline requires unique software development and resource allocation. In some cases, it has also been noticed that sequence depth coverage further increases if there are more numbers of sequencing errors in the reads. Therefore, before starting the assembly, it is imperative to collect the information on genome size, rate of sequencing errors introduced by the given NGS, repeat content, and the degree of genome duplications.

As there is limited such information for the majority of the crop plants, it is suggested to start the assembly process with a small pilot study using single-end or short-insert sequencing, and the genome size, degree of heterozygosity, and quality of the sequence assembly can be approximated using a k-mer counting approach (http://josephryan.github.com/estimate_genome_

size.pl). To this end, Seqanswers, a web forum, provides procedures and other information that is required for the analysis and interpretation of k-mer counts. As a general rule, if the target genome contains a high degree of repetition and genome duplications, it is advisable to have a long read length either through TGS or preparing long-insert data for perfect assembly.

Even if long reads are generated for the assembly process, specifically tailored bioinformatics pipelines are required as heterozygous regions of the genome have an undesirable impact on the assembly and polyploid species require additional efforts during genome assembly.

Such challenges can be avoided if the WGS program involves true inbreds of the given species or parthenogenetic or gynogenetic off springs if it exists. It is also imperative to collect other metadata information (including age and sex of the individual, time, and exact place of sampling besides which tissue (leaf or flower or seed) was collected for genome sequencing), which would be useful for assembly and further comparison with similar WGS programs.

Isolation of high quality, intact, non-degraded, and ~1 mg of DNA of the target plant species is another prime factor that decides the success of WGS and determines the quality of long-insert size libraries. Before starting the WGS process, the integrity of the DNA is required to be verified with high-resolution gel electrophoresis such as pulse-field gel electrophoresis, and the DNA fragments of >100 kb are preferably used for WGS. It is also imperative to validate the quality of the DNA library preparation and to ensure it equally represents different parts of the genome. Thus, it is suggested to have more than one library (replicates of each size class) to represent the same WGS program.

It should also be noted that few of bioinformatics programs used for genome assembly (such as ALLPATHS-LG) require input data as a predefined mix of sequencing libraries, and it is also required to fix the read orientation (which is again fixed by the type of NGS platform employed in the sequencing; for example, reads can face inwards (e.g., Illumina paired-end sequencing) or outward (e.g., Illumina mate-pair sequencing) with respect to the actual DNA sequence of the target organism.

Wrongly oriented reads will also arise due to sequencing of pairs from within the original DNA fragment instead of ends or due to generation of mate-pairs with unusual insert sizes (which lead to chimeric sequences from nonadjacent genomic regions). Therefore, these artifacts are required to be trimmed off and only the unique reads with adequate quality should be employed for genome assembly.

During the validation process of sequence assembly, care should be bestowed on duplicates as the majority of the NGS technologies involves PCR amplification for the sequencing. PCR introduces significant numbers of duplicate reads; when there is a length distribution it will lead to artifacts during genome assembly and it is difficult to differentiate the actual gene duplicates in the target genome and the artifacts. Therefore, if the duplicated reads are considered during the assembly (which constitute considerable percentages of short-insert size libraries, ~10 kb), it may impair the quality validation of genome coverage and it is better to avoid such duplications during the assembly process.

Another key point during the genome assembly is the size of the reads that are used to decide the sequencing depth. As a thumb rule, having a fine mix of contig read lengths in the range of 0.2–40 kb significantly increases the depth coverage. In general, contig size or long insert size of >20 kb provides adequate final contiguity and scaffold size of the genome assembly with improved quality.

It is also imperative to arrange for adequate data storage and management facility before the start of the project, as the enormous amount of sequencing data would be generated even with minimum depth coverage. For example, several hundred gigabytes of sequence data would be generated if the WGS project aims to cover 100X genome, and hence the temporary files that were generated during the genome assembly will require terabytes. Therefore, connecting with local or national computing grids or cloud storage facilities and establishing a strong association between bioinformatics and biological researchers in dealing with such huge data would be essential for a genome assembly program.

Instead of utilizing the readymade bioinformatics pipelines that were used for genome assembly in the model plant species, it is advisable to design a unique pipeline for the given plant genome as the conservation strategy is not optimized in all the cases, as highlighted above. Further, ensuring the expertise to develop, execute, and curate bioinformatics pipelines for the genome assembly in the core research group is also essential. The core research group should also be acquainted with open-source operating systems such as UNIX and writing and applying scripts of Perl or Python languages since the majority of the genome assembly programs have been developed under UNIX environments.

Assessment of sequence data quality, GC content, repeat abundance, and amount of duplication should be done before starting the genome assembly process, for which tools such as FastQC (http://www.bioinformatics.babraham.ac.uk/projects/fastqc) can be employed. The use of FastQC offers simple ways to provide summary statistics which are considered as a good starting point for genome assembly.

Alternatively, ConDeTri can be used to remove low-quality data and reads resulting from PCR duplications. Similarly, the employment of a stand-alone error-correcting program using a k-mer count approach (such as SOAP*denovo* pipeline) can also be tried. Some of the assemblers, such as ALLPATHS-LG, perform both trimming and error removal within the assembly pipeline using the raw reads. Nevertheless, as stated earlier, such an assembling strategy should be employed based on the need of the individual WGS project.

It is also important to remove primer and vector sequences (that were used during library preparation and which can be removed effortlessly using programs such as *cutadapt*) and other unrelated sequences (for example, DNA sequences of PhiX phage is used to calibrate sequence quality scores in Illumina sequencing). Failing to do so will increase the percentage of contaminant sequences (chimeric and contaminated contigs) and finally affect the authenticity of the assembly process. The use of a short read aligner such as BWA will easily remove such vector contamination sequences.

As and when a new challenging situation arises due to different datasets, large arrays of programs have been developed for de novo assembly of shotgun whole-genome sequencing data, and the list is ever-growing. Some of the algorithms reduced the dis-assembly, whereas others focused on the contiguity even at the cost of perfection. Each program used in genome assembly differs in terms of speed, scalability, and the quality of the final genome sequence, and hence it is tricky to select a universal assembly program. This is attributed to the fact that the target genome has its unique structure, size, base composition, repeat content, and polymorphism level.

At this point, it is suggested to have an open discussion with the experienced program developers before designing the own assembly pipelines besides collecting the required information from the literature and websites of assembly software.

The most commonly used genome assembly approach is overlap-layout-consensus (OLC), and it is very common to employ OLC for long read assembly that is resulted from Sanger sequencing (which uses Celera assembler, Arachne, and PCAP) or Roche 454 sequencing (which uses Newbler). However, for short-sequence data obtained from Illumina or SOLiD, OLC has been found to be computationally intensive. Despite the excessive runtime requirement, some of the short-read assemblers (for example, Edena, SGA, and FERMI) employ OLC.

However, the majority of methods that are involved in de novo assembly of short reads uses either extension-based methods or De Bruijn (or Eulerian) graph algorithms. Though the extension-based assemblers (examples include SSAKE and JR-Assembler) have been noticed to be efficient in computation (with respect to requirements of both computational memory and time), they are found to be unproductive when they deal with sequencing errors, repeat regions, and high levels of nucleotide polymorphism.

To circumvent these inefficiencies, De Bruijn graphs are now widely employed in the assembly of short-read data. In this strategy, the short-reads are partitioned into k-mers (i.e., divided into substrings of the k read length sequences) which are subsequently form the nodes of the network (which form the graph) that are linked together when sharing a k-1 mer. Almost all the extensively used assembly software (to name a few: SOAP*denovo*, ALLPATHS-LG, ABySS, and Velvet) employs De Bruijn graph algorithms for genome assembly.

An alternate to the k-mer frequency-based approaches, the expected genome size can also be obtained from C-value data. Assembly contiguity can also be evaluated using the N50 statistic (which is defined as 50% of the assembled nucleotides found in the contigs (contig N50) or scaffolds (scaffold N50) are having a minimum of the given length (expressed in bp or Kb or Mb). Therefore, the use of N50 statistic illustrates a kind of median of assembled sequence lengths; the higher the N50 length, the greater the long sequences assembled (refer to Table 10.1 to get knowledge on how different studies produced different contig and scaffold length in Moringa).

An update on N50 referred to as NG50 and 'NG Graph' is proposed recently by incorporating the expected genome size. It offers a better way of visualizing and appraising variations in contiguity among assemblies. At the same time, N50 and NG50 statistics should be used cautiously as it denotes only the contiguity and not the accuracy of the assembly. As stated earlier, information from remapped paired-end or mate-pair data by using an algorithm such as REAPR should be used to find sequence errors in the assembly. Publically available transcriptome sequencing data (RNA-seq data) and expressed sequence tags (ESTs) of the target species also constitute key additional resources for validating sequence accuracy and for correcting sequence errors in scaffolding. These resources would also be important to fill the gaps when gene sequences span across contigs.

Besides, some of the approaches (such as Atlas, Ray, and MaSuRCA) utilized 'hybrid' assembly algorithms by merging features of diverse algorithms and making use of data from several sequencing platforms. Thus, it would be desirable to employ as many assemblers as possible (as that of an iterative process) when doing a de novo assembly and evaluate the resulting assemblies with several parameters that lead to reaching a fine-tuned consensus sequences.

Recently, the 'Align-Layout-Consensus' algorithm has also been suggested for genome assembly in which the overlap stage of the de novo assembly is substituted by alignment of reads to an intimately associated reference genome, and contigs and scaffolds are then built using the data obtained from overlapping read. Thus, it would be computationally simple when compared with OLC.

At the end of the contig assembly, it would be common to notice a stretch of Ns that needs to be processed before the scaffolding step. Though this scaffolding assembly is inherently included in assembly programs, stand-alone algorithms (for example, SSPACE and BESST) are also available. Other algorithms (such as GapCloser, GapFiller, and iMAGE) can remove the gaps (due to the presence of N's) using the original read-pair information. Recently, long-read data obtained from TGS strategies have been employed to fill the N's in the scaffolds.

At this point, it should also be mentioned that there is some commercial software (such as CLC workbench or Lasergene from DNASTAR) that can also be used for genome assembly as it is more user-friendly and does not require any bioinformatics skill. However, it involves initial investment in purchasing the license, and almost it is impossible to inspect or alter the program, which is specifically needed for the given plant genome. Sometimes the vendors of DNA sequencers also supply assembly software along with the instrument.

10.2.5 Annotation

In order to realize the complete potentials of plant WGS project, it is essential to annotate those DNA sequences (refers to obtain meaningful data from the sequences) with biological information such as gene models, enumerating the roles of the predicted genes using gene ontology (GO) terms (by employing Gene Ontology Consortium or 'Kyoto encyclopedia of genes and genomes' (KEGG) pathways), identification and characterization of regulatory elements such as microRNA (by employing miRBase) and epigenetic modifications (using ENCODE Project Consortium).

A complete annotation process requires adequate efforts and bioinformatics skills, and annotation is found to be critical in non-model plant species. To begin with, annotation of non-model crop species often starts with the identification of protein-coding sequence (CDS) or transcripts, though it is a challenging task as there is no preexisting gene model.

The success of the annotation process largely depends on the quality of the assembled genome sequence, and it is desirable to start the annotation process with the contiguous near-complete ($\sim$90%) genomes that have less numbers of gaps. In general, it has been realized that the larger the genome size, the greater the gene models, and hence it is essential to have more contiguous assemblies before starting the annotation process.

There are two important phases in the genome annotation: the computational phase and the annotation phase. The computational phase consists of collecting and employing different types of molecular proofs from other genomes or from the target plant species-specific transcriptome data for the preliminary prediction of gene models. Whereas in the annotation phase, the gene models are predicted using a set of rules outlined by the annotation pipeline (which are usually based on start and stop site of the transcript or CDS, regulatory elements, exons and introns characteristics, *etc.*,).

In both phases, it is imperative to avoid or mask the repetitive sequences such as low-complexity regions and transposable elements of the target genome before starting the gene prediction process as repeats are less conserved. This can be done using RepeatMasker (freely available at http://www.repeatmasker.org) followed by RepeatModeler or RepeatExplorer which will help to construct a species-specific repeat library.

Consequent to the masking of repeats, the genome sequences can be subjected to *ab initio* algorithms (e.g., AUGUSTUS) that are trained on predicting gene models (or CDS) using the established genome annotation of related species. This gene model prediction process can also be complemented or confirmed with protein alignment algorithms (such as tblastx; freely available at NIH—NCBI website) and syntenic proteins found in other plant species.

Indeed, the best confirmation step for predicted gene models is the use of EST or RNA-seq data obtained from various tissues of the target plant species from which WGS data has been generated. Employing such data would not only provide information on CDS but also other relevant information on gene models such as splice sites, transcription start sites, and untranslated regions (UTRs). More particularly, strand-specific sequencing of mRNA can provide more resolution on the predicted gene models and offer efficient strategies for transcriptome assembly, which will ultimately help in the precise evaluation of the WGS data.

Consequent to the *ab initio* prediction and confirmation using protein-, EST-, or RNA-alignments and gene models have to be integrated into a final set of well-annotated genes. It is considered a herculean task, as the predictions from the above methods are mostly incomplete and occasionally contradicting with different gene models. Hence, the predicted gene models require meticulous manual curation. Though certain algorithms (such as MAKER and PASA) can automate this curation process using the molecular evidence from different sources, they require qualitative validation which is usually done by evaluating the open-reading frames and other components of the gene.

For example, our initial attempt in identifying the genes in the regional Moringa cultivar (but

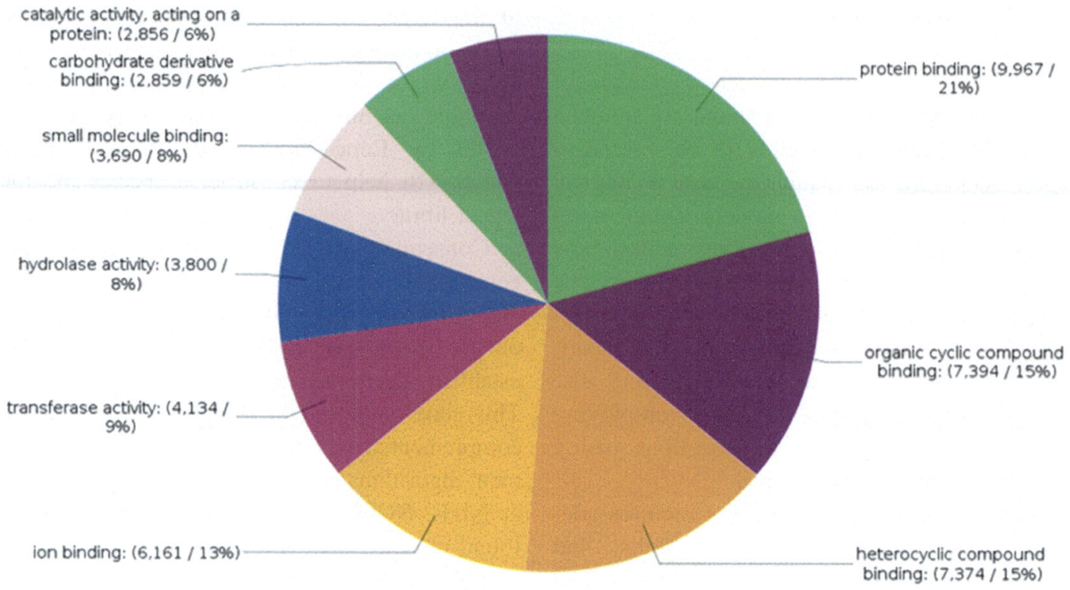

Fig. 10.2 Annotation of molecular functions of *M. oleifera* PKM1 genes

globally adapted), PKM1, has disclosed the following: among the 41,801 predicted gene sequences, 19,588 sequences retrieved blast hit along with gene ontology mappings such as genes involved in cellular processes, molecular functions, and biological process. The majority of genes were found to be distributed in the organelle (19%), intracellular organelle (19%), cytoplasm (17%), membrane (11%), and cell periphery (8%). Based on the motif and domain similarity search, annotation of 41,801 sequences using the Interproscan database was also done. It provided the information that PKM1 genes were involved in the cellular process (15%), metabolic process (13%), response to stimulus (10%), biological regulation (8%), regulation of biological process (8%), and developmental process (8%). Most of the genes were found to contain functions such as protein binding (21%), organic cyclic compound binding (15%), heterocyclic compound binding (15%), ion binding (13%), and transferase activity (9%) (Fig. 10.2).

In order to increase the authenticity of predicted gene models, it is vital to critically visualize the methodical issues such as intron leakage (which refers to the wrong prediction of introns instead of exons due to the presence of pre-mRNA) or gene fusion. This process can be done using WebApollo (a product of the GMOD project), which facilitates easy editing during annotation by utilizing a visual interface.

Owing to the tremendous development in NGC and TGS, draft genome sequences of several plant species are being developed and deposited in public databases such as ENSEMBL from the European Molecular Biology Labs (EMBL) and the Wellcome Trust Sanger Institute, or genomic databases from the National Center for Biotechnology Information (NCBI). Though it is not possible to annotate all the draft genomes, now it is possible to upload the draft sequence alone in NCBI as a BioProject, which allows other users to get better assembly and annotation.

10.2.6 Mining for Novel Genes and Regulatory Sequences in Moringa

Unraveling the imperative and complex plant molecular mechanisms, such as drought tolerance, pest, and disease resistance in plants, is a long-standing objective. Hence, the discovery of novel biochemical and cellular processes that provide a deeper molecular understanding of the above is the major goal of genomics approaches.

Genome mining refers to the discovery of novel genes, regulatory sequences, and metabolic pathways that lead to synthesize novel products using WGS and other omics data. Thus, genome mining offers a unique practice of identifying novel metabolic pathways that are involved in the assembly of structurally complex bioactive and medicinally important natural products. In addition, the availability of new and additional WGS and other omics data that are being accumulated in the public databases also provide opportunities to explore other untapped molecular information on the regulation of gene expression (such as cataloging novel regulatory elements like micro RNAs (miRNAs)).

Continuous advancements and progress in developing the algorithms to predict gene models in a high-throughput manner facilitate fast prediction and evaluation of modular biosynthetic assembly lines that are involved in useful metabolites production and confirming the novelty of the identified metabolite products with simple in silico strategies rather than using complex validation procedures that involve biological systems.

10.2.6.1 Genes that Code for Important Nutritional and Medicinal Properties in Moringa

In contrast to the traditional way of discovering novel plant drugs, genome sequencing and other omics tools offer another novel route to identify medicinally important phytocompounds from the plants including Moringa.

For example, Pasha et al. (2020) presented the combined transcriptome that was developed from leaf, root, stem, seed, and flower tissues of *M. oleifera* variety Bhagya and documented that there were several tissue-specific plentiful copies of mRNAs that are involved in medicinally imperative metabolites production. Particularly, they have shown transcriptomics evidence that were involved in the metabolism of quercetin, kaempferol, benzylamine, and ursolic/oleanolic acid that impart medicinal value to *M. oleifera*. Besides, they have also shown relative expressions of transcripts that were essential to synthesize vitamins A, C, and E and different levels of mRNA expression that were involved in metal ion transporting in diverse tissues.

With the help of the PlantCyc database, qRT-PCR assay, quantification of metabolites and minerals, and molecular phylogenetic analysis, Pasha et al. (2020) could able to report 36 candidate genes and stress-responsive transcription factors and provided evidence for well established metabolic pathways involved in the biosynthesis of four secondary metabolites (*viz*., quercetin, kaempferol, α-aminotoluene (popularly known as benzylamine), and ursolic/oleanolic acid).

Quercetin belongs to flavonoid groups, and it has been found to present in Moringa and other vegetables, fruits, and beverages. Quercetin has experimented with several cancer cells as a chemoprevention agent owing to its antioxidant, anti-tumor, and anti-inflammatory activity and has been concluded that little doses of quercetin were sufficient to inhibit the cell cycle arrest at the G_1 phase and thus prevent the proliferation of cancer cells.

In plants, 4-Coumarate-CoA ligase (4CL), Chalcone synthase (CHS), Chalcone flavone isomerase (CHI), Flavonone 3-hydroxylase (F3H), Flavonol synthase (FLS), Tricin synthase (OMT), and Flavonoid 3'-monooxygenase (F3'H) were found to be involved in the quercetin biosynthesis. Out of these seven enzymes, 4CL, CHS, CHI, F3H, and FLS were also involved in kaempferol biosynthesis.

Interestingly, Moringa transcriptome and cross-family connections reported by Pasha et al. (2020) provided evidence analogous to the above enzymes. Fascinatingly, F3H and FLS seemed to be encoded by the same transcript. According to them, FLS can act bifunctionally in other plants (i.e., it can also act as an F3H) and OMT can metabolize quercetin in certain plants using an unidentified mechanism. Differential expression analysis, qRT-PCR assays, and HPLC analysis performed by them have documented that identified Moringa transcripts codes for all the seven enzymes of the quercetin/kaempferol biosynthesis pathway were present in higher amounts in Moringa flower, leaf, and seed.

The addition of benzylamine in drinking water increases glucose tolerance and is found to diminish body weight gain and circulating cholesterol when fed with a high fat diet (Iffiú-Soltész et al. 2010). In bacteria, N-benzylcyanide is converted to benzylamine by the enzyme N-substituted formamide deformylase (NFD). In plants, biosynthesis of benzylamine involves the conversion of N-benzylformamide to benzylamine by NFD, and this enzyme has been well described in barley (*Hordeum vulgare*). Pasha et al. (2020) has made a sequence search for the NFD in the *M. oleifera* and found that there was a single homologous sequence hit. This was also validated by both phylogenetic investigation, qRT-PCR results, and mapping of functionally important residues. These experiments were concluded that NFD was highly expressed in Moringa seeds followed by leaf and stem tissues when compared with flower and root tissues. Another striking finding made by Pasha et al. (2020) using quantitative analysis of the metabolites was that dibenzylamine was found to be high in the root sample which denoted that benzylamine might be synthesized in other tissues and translocated to root for storage.

(3S)-2, 3-epoxy squalene oxide is a precursor for synthesizing both ursolic and oleanolic acids which will be converted to α-amyrin and β-amyrin, respectively, by the enzymes α-amyrin synthase and β-amyrin synthase. Cytochrome P450 (CYP450) (Amyrin monooxygenase) acts on α-amyrin and β-amyrin which ultimately lead to producing ursolic and oleanolic acid, respectively. Though CYP450 is a large superfamily (known to have at least 52 subfamilies), CYP450 716A/B subfamily was successfully established that it was involved in the biosynthesis of Ursolic acid and Oleanolic acid. Sequence searches of CYP450 domains in the *M. oleifera* have reported six transcripts in Moringa and among them, two were noticed to cluster with the CYP450 716A/B subfamily Pasha et al. (2020). They have also demonstrated that these two transcripts were highly upregulated in Moringa root tissues. At this point, it would be appropriate to recollect that the roots of *M. oleifera* and these two secondary metabolites have long been utilized as anti-infertility agents.

A similar kind of analysis of transcriptomic data of *Moringa oleifera* cultivar PKM1 has been done at this laboratory and such effort has identified several other pathways that lead to synthesize the important medicinal compounds such as naringenin, apiforol, apigenin, and isorhamnetin (Boopathi et al., in preparation).

10.2.6.2 miRNA Genes and Their Targets

All the eukaryotes possess a vital conserved gene expression regulatory system, which is mediated by small RNAs (sRNAs). Plants are also having such sRNA-mediated gene regulation. These sRNAs are derived from long double-stranded RNA (dsRNA) by RNase III, and it generates two major sRNAs: small interfering RNAs (siRNAs) and microRNAs (miRNAs).

In general, miRNAs are small in size (~21–24 nucleotides) and produced endogenously. Though they are not coding for any proteins, they have negative post-transcriptional regulation of gene expression by silencing the target mRNAs. All the biological processes such as development, growth, and stress responses are known to be negatively affected by these miRNAs, and hence they are referred to as negative regulators of target gene expressions.

Plant miRNA-mediated silencing is mainly mediated by the robust complementarities between the miRNAs and their target mRNAs, which is the key region that favors the endonucleolytic cleavage. It has been established (with *Arabidopsis* mutants, which were defective in miRNA action) that miRNA–guided silencing can also be mediated with translational inhibition, which is different from endonucleolytic cleavage.

Though plant miRNAs have been investigated for a long time and the information on the biogenesis of miRNA has been disclosed in detail, the majority of other molecular aspects (such as their origins and function) are yet to be resolved. Such information would be more beneficial to design a successful experiment that focuses on desirable gene expression and genetic improvement of crop plants. Further, those results will also be useful to understand the evolutionary events at the molecular level which would be supportive evidence for phenotypic evolution. For example, the identification of miRNAs involved in the adaptation of plants to different ecosystems and climate vagaries will enable designing efficient molecular breeding strategies.

Widespread documentations of miRNAs in major crops are available but Moringa specific, conserved and unique miRNAs and their mRNA targets are limited. The first and only report (as of date) on Moringa miRNA was provided by Pirrò et al. (2016) using high-throughput sequencing of a small RNA library constructed from *M. oleifera* seed. They identified 94 conserved and two novel miRNAs (which were further validated by qRT-PCR assays) from 31,290,964 raw reads. In addition to establishing their value in gene regulation, they also described the pharmacological potential properties of those miRNAs.

For example, few of the predicted Moringa miRNAs have shown functional homology to human miRNAs which can regulate the human genes when they were transfected into cell lines. Besides, their bioinformatics results enabled them to offer novel insights into a potential cross-species control of human gene expression. Thus, such miRNA research in Moringa will open up new avenues in improving human health by the transfer of candidate miRNAs from this valuable species.

Such effort in cross-species miRNA transfer is not new, and it has already been shown using rice and mice. Zhang et al. (2012) successfully shown that *osa-miR-168a* and other exogenous miRNAs of rice can be acquired by mice through food intake. It has also been demonstrated for the first time using *in vitro* and *in vivo* functional studies that cross-kingdom gene regulation can occur as exogenous miRNAs inhibit mammalian gene expression in the liver (Mlotshwa et al. 2015). Similarly, orally administered cocktails of endogenous tumor suppressor miRNAs (which mimics the characteristics of plant miRNAs) have decreased the tumor burden in mice (Shu et al. 2015).

The aforesaid experiments proved that plant-derived miRNAs can be absorbed by the mammalian digestive tract, and those absorbed miRNAs can regulate mammalian gene expressions. Further, engineering those edible and useful miRNAs will enable efficient suppression of mammalian tumors (or other target traits) which will be an effective, nontoxic, and inexpensive chemo-preventive treatment for cancer.

Thus, Moringa miRNA research in these directions would offer huge benefits to the health industry as it is substantially demonstrated that Moringa has a large array of medicinal properties, which may be partly due to the presence of such miRNAs.

10.2.7 Applications of Genomics in Moringa Genetic Improvement

The key application of Moringa genomics research is the identification of large arrays of molecular markers and novel genes associated with agronomic and economic importance. These markers/genes provide valuable resources to unravel the molecular mechanisms of expressing and regulating imperative phenotypes, which will be directly used to precisely and rapidly breed the new Moringa cultivars that meet the stakeholders'demand.

There are at least four references and resequenced Moringa genome sequences (Table 10.1) which enables swift discovery of potential candidate genes using different bioinformatics strategies (see Sect. 10.2) and discovery of single nucleotide polymorphism (SNP), Insertion-Deletions (InDels), simple sequence repeats (SSR), and other marker classes through comparison of the reference sequence with resequencing data obtained from different cultivars. For example, more than twenty thousand of SSR markers have been identified and characterized using the genome resequence data obtained from Moringa cultivar PKM1 at this laboratory (see Sect. 9.2.1).

In general, the discovery of novel genes depends on unassembled genome sequence data and expressed sequence tags (EST), if there no reference sequence is available. Though such data can also be employed SNP and SSR marker discovery, it requires genetic mapping to pinpoint these markers on the genome (see Chapter 9). Such genetically mapped molecular markers offer potential applications in the Moringa breeding program for its inherent improvement through genetic diversity analysis, cultivar identification, phylogenetic analysis, and characterization of genetic resources association.

Instead of conducting labor and resource-intensive field and glasshouse screening, molecular markers can be used to screen the genetically characterized traits as tracking the inheritance of several agronomic traits is found to be complex using field and greenhouse experiments. For example, molecular markers that are found to be associated with underlying genetic determinants of disease resistance and complex traits such as drought tolerance and yield can be used to select for these traits, which increases the efficiency of crop breeding.

Thus, genomics research in Moringa has widespread applications which ultimately improve the efficiency of the strategies that focus on genetic improvement of Moringa for valuable agronomic and economic traits. However, progress toward this end is limited, and this laboratory is being involved in bringing the potentials of genomics results into regular molecular breeding approaches.

10.3 Comparative Genomics in Moringa: Approaches and Applications

Availability of genome sequences of different related and unrelated crop species allows detailed structural and functional comparisons of genes and regulatory elements involved in various plant growth and development processes and biotic and abiotic stress responses. Such efforts have functional utility in unraveling the molecular process vital plant developmental phenomena such as reproductive development and defense signaling and evaluate the same phenomena in different plant species by studying the functions of orthologous genes (see below). Extensive studies on orthologous genes in different crop species have demonstrated that orthologous genes exhibited conflicting functions and such variations in functions and regulation of expression of those orthologous genes formed the basis for diversity among the investigated crops. The key application of these comparative genomics studies is such information enables a deeper understanding of the evolution of those crop plants and precise transferring of potential candidate genes from one plant species to the other.

10.3.1 General Approaches Employed in Comparative Genomics

A large number of comparative genomics tools and pipelines have been developed for comprehensive exploitation of the genome sequence data obtained from different crops at systems-level frameworks. Analytic and visualization tools have also been designed to have an efficient comparative analysis of coding, non-coding, and small molecules across plants.

One such tool is Gramene (http://www.gramene.org), which was designed to perform comparative functional analyses of genomic data obtained from both model and non-model plants, and it has been found to be useful for several global agricultural-biotechnology researche. The key interface of Gramene is the Ensembl Genome Browser that facilitates visualization of molecular variation, gene expression, and epigenomic modifications across the sequenced plant species. Besides, it contains a BioCyc-based platform that facilitates the building of cellular-level metabolic pathway models. Plant Reactome portal (https://plantreactome.gramene.org) is another provision in Gramene that provides improved visualization of metabolic pathways at the sub-cellular level and illustrates various molecular events, macromolecular interactions, genetic regulatory networks and metabolic pathways, and network that occurs during key plant growth and development processes.

Besides, each result of the Gramene's pipeline and Plant Reactome pathways are linked to related public genomic resources which enable additional information. For example, the identification of agronomically important genes displays real time basal tissue-specific gene-expression profiles and anatomogram images that are fetched from the EMBL-EBI's Expression Atlas.

Gramene's powerful and flexible document-based architecture provides easy access and analysis of the data across its portals (ENSEMBL, Reactome, and Expression Atlas), which provides interactive and graphically summarized views on gene features, genomic physical position, phylogenetic trees, gene expression profiles, pathways, and cross-references. Gramene also supports species-specific Genome and Pathway Browsers, and all the analyzed data can be bulk downloaded both in graphical and tabular formats.

According to the recent publication (Tello-Ruiz et al. 2020), Gramene now shelters 93 plant genome browsers, gene-Orthology based pathway projections for 107 plant species (hosted from both Gramene Genome portal and other sources), and gene-expression data in the Expression Atlas from 941 experiments conducted with 28 plant species. Domain Informational Vocabulary Extraction (DIVE) algorithm is a recent addition to Gramene which can be used to automatically extract gene functional information.

10.3.2 Comparative Genomics

10.3.2.1 Within Moringa

In line with recent advances in 'omics' methods including proteomic, transcriptomic, and metabolomic methods, Moringa tissues have also been explored and provided a deeper understanding of molecular aspects of the different developmental processes such as seed germination and secondary metabolite synthesis in different parts of Moringa under different spatial and temporal conditions.

For example, changing patterns of protein expression between dry (matured) and germinating *M. oleifera* seeds were documented by Wang et al. (2020) by employing label-free quantitative proteomics. This study has particularly reported a peptidase and three different endopeptidases that were expressed during Moringa seed germination. Properties of these peptidases along with their enzymatic activity have been investigated by analyzing the amino acid composition and determining proteolytic and milk-clotting activities. Such characterization studies on proteases obtained from *M. oleifera* germinating seeds have potential applications in the food industry which produces functional foods with bioactive peptides and processes cheese with milk-clotting agents.

As stated before, Pasha et al. (2020) employed both transcriptomics and metabolomics to unravel the molecular basis of differential biosynthesis of medicinally imperative secondary metabolites using five different tissues (leaf, root, stem, seed, and flower) of *M. oleifera* cultivar Bhagya. Besides providing a list of up- and/or down-regulated genes that were involved in particular metabolite synthesis in each tissue, this study has also offered experimental proof for selecting appropriate tissue that has adequate synthesis of target metabolites that possess medicinal value.

10.3.2.2 Moringa and other Crop Genomes

An attempt on comparative genome analysis of *M. oleifera* and four woody plant genomes belongs to the *Dicotyledons* clade (namely, *Vitis vinifera*, *Cajanus cajan*, *Carica papaya,* and *Malus domestica*) by Tian et al. (2015) have provided preliminary molecular evidence for the evolution of these wood species by constructing a phylogenetic tree.

The results of phylogenetic analysis have clustered the *M. oleifera* close to *Carica papaya* which promoted to classify *M. oleifera* in Brassicales. Subsequent phylogenetic and whole-genome duplication (WGD) analysis of four Brassicales species namely *Arabidopsis thaliana*, *Brassica rapa*, *Carica papaya*, and *Moringa oleifera* suggested that there were several rounds of WGD events in Brassicales and provided a prelude on the evolution history of Brassicales.

This study also helped to identify several species-specific gene families and depicted the list of positively selected genes in *M. oleifera* that lead to impart unique characteristics such as *M. oleifera's* high protein content, fast growth, and heat and other abiotic stress tolerance.

Orthologous Genes

The genes that exist within or between the species and share common features are called homologous genes. In general, homologs are genes that share sequence identity. Though the term homolog does not provide any information on evolutionary history, homology between genes generally offers some evolutionary relationship.

There are two groups of homologous genes: Orthologous genes and Paralogous genes. Genes identified in different species that are vertically originated from a common ancestor but codes for the same protein with the same enzymatic activity are referred to as orthologous genes (for example, the human α-globin gene and chimpanzee α-globin gene are orthologous). On the other hand, paralogous genes are homologous genes that are identified in the same species. Paralogous genes have evolved due to duplication events and code for a protein with similar but not identical enzymatic or other regulatory functions (for human α and β-globin genes are paralogous).

Identification of homologous genes across the plant species is generally accomplished by using homology searches using simple Basic Local Alignment Search Tool (BLAST) available at https://blast.ncbi.nlm.nih.gov/Blast.cgi?PAGE_TYPE=BlastSearch. Orthologous genes are the most widely searched gene groups which can be effortlessly identified for single-copy genes heuristic algorithms. Usually, orthologous genes are due to syntenic between species, sharing common flanking regions whereas among the paralogous genes synteny is lost.

In order to identify the orthologous genes in Moringa, Tian et al. (2015) have initially utilized BLAST to align the coding sequences of *M. oleifera*, *Vitis vinifera*, *Malus domstica*, and *Carica papaya*. The resulting 5,601 orthologous gene pairs were again analyzed by employing blasts, Ka/Ks_ Calculator 1.2, and MUSCLE and finally visualized the aligned orthologous genes using ClustalX. They have reported 566, 399 and 112 orthologous genes when *M. oleifera* was compared with *Carica papaya, Vitis vinifera*, and *Malus domstica*, respectively. They also provided additional molecular evidence to claim that higher production efficiency of *M. oleifera* protein synthesis machinery might be evolved recently.

Our preliminary attempt on finding orthologous genes has also revealed more closeness of the genes identified in *Moringa oliefera* cultivar PKM 1 with *Vitis vinifera* and *Carica papaya* (Fig. 10.3).

Sequence Divergence in Moringa Genomes

Recent advances in developmental biology and comparative genomics have clearly pointed out that physiological, as well as morphological characters, are chiefly regulated by multigene families. Therefore, further investigations on the evolution of multigene families are imperative to unravel the evolution of agronomically important phenotypes and significant strength of the relationship between the evolutionary and functional changes within multigene families has been reported.

Even though the numbers of the gene(s) that are gained or lost during evolution significantly vary among multigene families, they are continuously under birth and death evolution. Those multigene families, which were undergone several rapid birth and death evolution, usually have variations in the number of genes between closely related species or even between individuals of the same species. Though the reasons for these variations in gene number in multigene families across the species or individuals are not clearly pointed out, it is believed that initially the number of genes in a multigene family is decided by their functional requirement and later when it

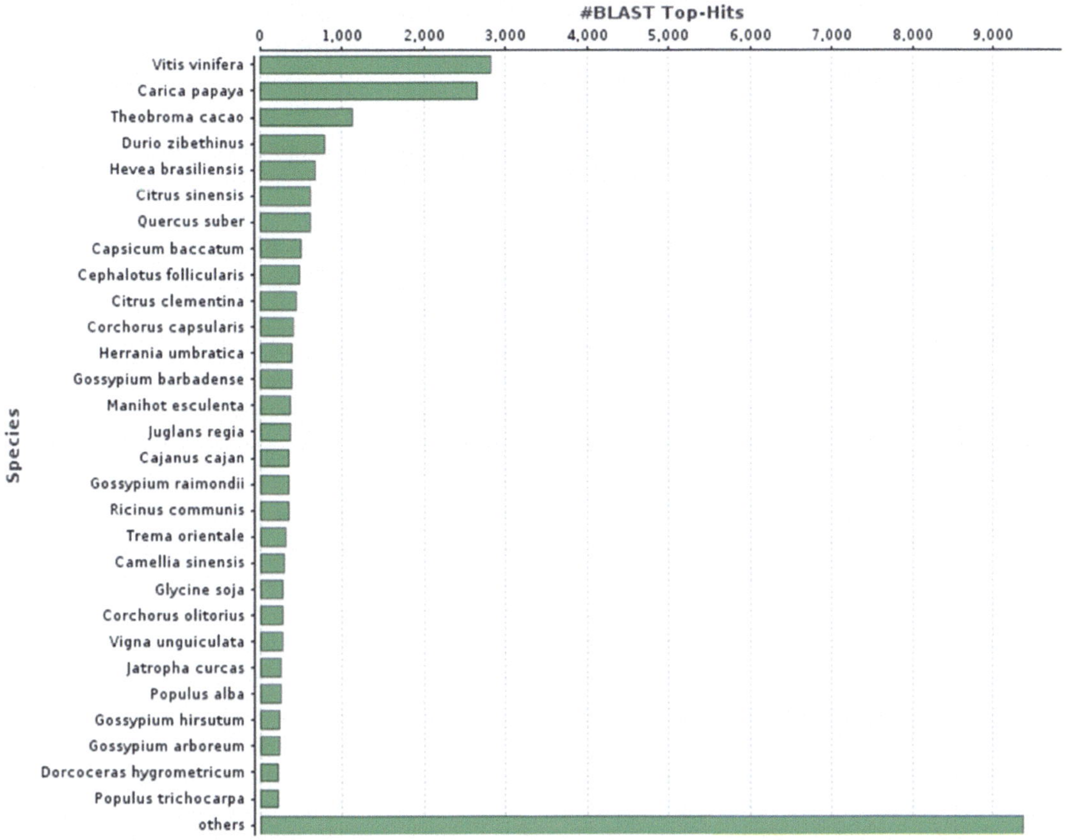

Fig. 10.3 Top BLAST hits of the *Moringa oliefera* cultivar PKM 1 with orthologous genes in other plants

reached a required level, the number of genes may increase or decrease by chance.

Availability of WGS on different crops has enabled genome-wide comparisons of numbers of genes, their patterns of variation, and degree of expansion within the crop superfamily. With the help of different molecular data, it has been established that divergence between monocots and eudicots might have occurred ~ 145 to 100 million years ago. Thus, combined studies will enable to clarify historical patterns of the evolution that lead to generating variations in multigene families. One such study was conducted by Xu et al. (2009) using 1,808 F-box genes from Arabidopsis, poplar and rice and concluded that there were 42 multigene families comprising genes that code for proteins with the same or similar domain organizations. Further, it has been documented that unraveling the history of the evolution of C-terminal domains might be complex but phylogenetic analyses will enable the identification of evolutionarily conservative and divergent genes.

Another pattern of the multigene family also occurs: quickly duplicating genes exist side-by-side with conservative genes which suggests that there was an unequal degree of gene duplication. Therefore, birth rates may be estimated on a higher side in some lineage-specific clusters than in the multigene family. This has also been noticed in well-known rapidly duplicating genes such as receptor-like kinase genes, disease resistance (R) loci, SKP1-like genes, and Type I MADS-box genes. Such observations have resulted to investigate the basic question: why do plants have evolved to possess so many duplicated genes that too in the rapid rate?

Chang et al. (2019) have used the protein and nucleotide sequences of Moringa and other plant species (namely *A. thaliana, Carica papaya, Citrus sinensis, G. max, M. truncatula, O. sativa, P. vulgaris, S. bicolor*, and *Theobroma cacao*) and constructed gene families by employing OrthoMCL software. This study has documented the clustering of 8,184 gene families of *S. birrea, M. oleifera, C. papaya, C. sinensis,* and *T. cacao,* and among them, 365 gene families (comprising 798 genes involved in fatty acid and diterpenoid biosynthesis and with cyanoamino acid metabolism) was specific to *M. oleifera.*

Other attempts in the characterization of gene families in Moringa have also provided some interesting clues on gene families and their regulatory roles though there were no evolutionary relationships with other related species (probably not due to the non-availability of genomic resources in the related species). For example, Pasha et al. (2020) have identified *M. oleifera* specific 2326 Transcription Factors (TFs) and grouped them into 96 families. By comparing these families with other related species (such as *A. thaliana, C. papaya, T. cacao,* and *O. sativa*), they have identified that significant numbers of Moringa-specific TFs were closely related to *C. papaya.*

It has long been known that *M. oleifera* is a hardy crop and has several drought-resistant properties. In order to provide molecular evidence for these properties, Pasha et al. (2020) have documented TFs specific to multiple biotic and abiotic stress responses such as C2H2, AP2-EREBP, NAM, WRKY, ARF, bHLH, MYB, Homeobox, bZIP, and HSF, and these ten TFs were occupying 32% of total TFs identified in their study. Among them, the C2H2 TF family, involved in defense mechanism and multiple abiotic stresses, was found to be the highest in number, followed by GRAS, MADS, and C3H. Thus, this study provided valuable candidate TFs for further investigations that have promising regulatory roles in biotic and abiotic stress resistance improvement in *M. oleifera* and other crops.

10.4 Moringa Organellar Genomes: Importance and Applications

Accumulating DNA sequence evidence have unambiguously demonstrated that plant mitochondria and plastids have originated as bacterial endosymbionts. Particularly, eukaryotic organelles have shown to have the eubacteria (*Bacteria sensu*) as the closest living relatives and the plastids might be originated from the blue-green algae (*Cyanobacteria*) whereas mitochondria might be originated from the α-subdivision of the purple bacteria (Proteobacteria).

Identification of structurally homologous group I introns and similar insertion sites in metaphyte, chlorophyte, and fungal mitochondrial genomes indicated that organelle genome might originate only once (monophyletic origin) *via* the primary endosymbiotic pathway. Nevertheless, some of the algal species have plastids that were acquired secondarily *via* a eukaryotic (instead of prokaryotic) endosymbiotic pathway.

Advances in molecular techniques enabled to complete sequence the organellar genomes though there are some limitations in annotations of these sequences. To circumvent these limitations, several algorithms specific to annotate the organellar genomes have been developed (such as Mitofy, CpGAVAS, Verdant, DOGMA, and GeSeq).

Among them, an online annotation tool, specifically developed for plant organellar sequences, is GeSeq (https://chlorobox.mpimp-golm.mpg.de/geseq.html), which perform high-quality annotations with customizable execution of batch files within a short runtime. GeSeq's functions provide all the facilities required for annotation and are equipped with reference sequences and an equivalent database for annotation of chloroplast-specific protein and rRNA-coding genes. If there is an appropriate reference set, GeSeq can also be used to annotate mammalian mitochondrial genome sequences.

Organelle genomes are best suited for evolutionary studies as they are inherited uniparentally and usually, there is no recombination. On the other hand, due to the poor degree of substitutions in the chloroplast genome and almost nil level of recombination in mitochondrial sequences, identification of desirable polymorphism among the individuals of the population is limited.

However, microsatellites or SSRs identified from the chloroplast genome shown to be useful on several occasions including in exploring intraspecific variability (see Sect. 10.4.3). SSRs derived from both chloroplast and mitochondrial genomes have revealed the parental lineage and demonstrated their roles in breeding and evolutionary genetics.

Recently, systematic analyses of population structure, plant ecology, and phylogenetic examination have employed annotated organellar genome sequences which enabled deep perceptions of those genetic features and provided novel information to molecular breeding of crop plants.

10.4.1 Approaches Used for Moringa Organellar Genome Assembly and Annotation

There are few attempts that focused on sequencing of the chloroplast genome of Moringa (Liu et al. 2019 and Lin et al. 2019). Identification and utilization of SSRs derived from chloroplast can be used to capture haplotypic variation and to describe the natural hybridization besides enlightening genetic discontinuity among Moringa ecotypes.

The first report on the complete chloroplast genome in Moringa was provided by Liu and his coworkers in 2019. By employing Illumina Hiseq 2000 Platform, Liu et al. (2019) have sequenced *M. oleifera* grown in Yunnan Institute of Tropical Crops (YITC), Jinghong, China, and obtained a total of 3.8 Gbp reads and assembled into a complete chloroplast genome. Dual Organelle GenoMe Annotator (DOGMA) was used to annotate the assembled chloroplast genome and deposited in Genbank (accession number of MH778650).

The above reported Moringa chloroplast genome contains 160,599 bp DNA sequences. Among them, 61.1% are AT and it classically forms a quadripartite structure. It contains 88,933, 19,482, and 26,092 bp long large single copy, small single copy, and inverted repeat regions, respectively. It was further reported that the Moringa chloroplast genome comprises 113 full-length genes which comprise 79 protein-coding genes, 30 tRNA genes, and four rRNA genes.

In order to evaluate the phylogenetic relationship with other 16 Brassicales taxa, Liu et al. (2019) have also constructed a maximum likelihood tree using RAxML after aligning the sequences with MAFFT. It was found that genome organization, gene content, and gene relative positions were almost identical among the investigated genomes and *M. oleifera* was clustered with the *Alyssum gmelinii*.

Similarly, Lin et al. (2019) have used Illumina Hiseq X to generate sequence data of ~5 Gb from *M. oleifera* maintained at the nursery Guangzhou, China. NOVOPlasty was employed to assemble the chloroplast genome and annotated with Dual Organellar GenoMe Annotator (DOGMA) as well as Geneious version 10.1 (http://www.geneious.com). Both the assembly and annotation process was verified and amended by evaluating with other plant species sequences deposited in GenBank.

Lin et al. (2019) reported 160,600 bp of the complete chloroplast genome sequence of *M. oleifera* with 36.78% GC content and deposited in GenBank (accession MH939149). It was also found that there were 88,577 bp long large single copy and 18,883 bp long small single-copy regions interspersed with two inverted repeats of 26,570 bp each. It was further predicted that there were a total of 131 genes in Moringa chloroplast genome (comprising 87 protein-coding genes, 36 tRNA genes, and 8 rRNA genes). In addition, HomBlocks, MAFFT version 7.307, and RAxML tools were used for phylogenetic analysis of the identified homologous blocks of Moringa chloroplast genome sequences with 12 other species from Akaniaceae, Caricaceae, Capparaceae, Cleomaceae, and Brassicaceae along with

Arabidopsis arenosa as an outgroup. The resulted maximum-likelihood tree has shown that *M. oleifera* was phylogenetically close to *Carica papaya*.

10.4.2 Understanding Evolution of Moringa Organellar Genomes

Comparative genome analysis also helps to identify novel genes that have evolved to adapt to the given environment and required other agronomical features. Such enlargement gene families maytake place de novo from DNA sequences that were actually non-genic or due to movement of transposable elements or horizontal gene transfer or during gene and genome duplication.

Whole-genome duplications (WGDs) and small-scale duplications (SSDs), especially tandem duplications, are the critical source of evolving unique genes that possess novel biological functions required for adaptive traits. WGD result into duplication of every gene in the genome and occur rarely, whereas SSD is duplication of only few genes and occurs continuously.

WGD events may undergo rediploidization and fractionations which lead to rearrangement of the genome and reduce the overall genome size and ultimately bring back the original diploid state. The majority of the duplicated genes in SSD have found to be lethal, and they will be either become pseudogenes or eliminated from the genome. Occasionally, few duplicated genes are preserved for a long time during evolution by acquiring special functions.

Though, multigene families and their expansion and contraction have been well explored in nuclear genomes, little has been explored in the organellar genome. One study conducted by Ojeda-López et al. (2020) investigated gene family turnover of orthogroups in Moringa and 10 other plant genomes using the Maximum Likelihood framework and found that there were different groups of gene families with enriched chloroplast related functions. A major portion of these gene family expansions was due to frequent transfer of lengthy plastid DNA or genes to the nuclear genome, and it was not due to WGD or SSD.

Further, Ojeda-López et al. (2020) have reported that though there was less number of genes in the relatively smaller Moringa genome, 101 gene families (comprising 957 genes), were found to possess considerable expansion and they were particularly involved to improve chloroplastic and photosynthetic functions. Interestingly, nearly half of the genes of Moringa expanded families were categorized as orthologs of *Arabidopsis thaliana* plastid-encoded multigene families.

It has also been pinpointed that the impact of plastid DNA on the structure and function of Moringa nuclear genes and genomes by investigating microsynteny and patterns of synonymous substitutions rates (Ojeda-López et al. 2020). Further, the majority of the duplicated genes in plastids were originated in recent times owing to the introduction of large plastid DNA segments (besides large numbers of short insertions) into the nuclear genome. Thus, a large portion of plastid DNA in the Moringa nuclear genome (4.71%, the largest reported so far) documented in this study highlighted that the chloroplast genome also contributed to the evolution of multigene families in the nuclear genome.

Ojeda-López et al. (2020) also identified expansions and contractions in several Moringa-specific gene families involved in secondary metabolism (particularly in the glucosinolates metabolic pathway which is a novel bioactive natural compound found in Moringa). Glucomoringin and glucosoonjnain are the two major glucosinolates reported in Moringa, and glucosoonjnain seems to impart a bitter harsh taste of Moringa leaves. It should be noted Moringa ecotypes that are widely cultivated are rich in glucomoringin, whereas the wild Moringa types possess higher glucosoonjnain. This simply indicates that the domestication of Moringa has focused mainly on the selection of better taste variants that has less glucosoonjnain.

Owing to the richness of glucosinolates and their derivatives such as isothiocyanates (which are due to hydrolysis of glucosinolates by myrosinase), Moringa was considered to encompass several medicinal properties (see Chap. 4). When compared to wild Moringa types, less level of myrosinase activity was reported in domesticated Moringa. However, the myrosinase multigene family was found to be, although not extensively, expanded in Moringa (Ojeda-López et al. 2020) which opened up new genetic and biochemical research questions in bioactive compound production in Moringa. Such investigations will enable us to identify potential genes and gene families involved in the expression of important biological, agronomical, or pharmaceutical properties besides providing valuable information on the evolution of nuclear genome structure and function.

10.4.3 Development of Organellar Genome Specific Molecular Markers

In order to investigate population genetic structure and phylogenetic relationships, both organellar genome sequences are being increasingly used as they display different patterns of genetic differentiation compared to the nuclear genome attributing to their uniparental mode of transmission.

Thus, employment of all three (nuclear, plastid, and mitochondrial) genomes would provide a complete picture of plant population differentiation and evolution. In view of that, the development of molecular markers, especially simple sequence repeats (SSRs or microsatellites) from chloroplast and mitochondrial sequences would increase the efficiency of such genetic studies.

Available accumulating evidence exhibited that chloroplast DNA contains SSRs that are relatively short and several mononucleotide stretches such as $(A)_9$ or $(T)_9$. They are ubiquitous and generate high polymorphic information content. It has been noticed in some taxa that nuclear markers cannot disclose genetic discontinuities and distinctiveness as interbreeding has reduced the genetic evidence of past demographic patterns. On the other hand, chloroplast-derived SSRs have clearly revealed the genetic differences among or between plant taxa even with slight morphological differentiation.

Further, investigations on comparative genomics and phylogenetic relationships can be effectively sampled due to the conservation and homology of chloroplast genome sequences across the plant kingdom which has diverged millions of years ago. Thus, plastid SSRs are now recognized as high-resolution markers as they provide a large amount of detail to

i. Examine the patterns of cytoplasmic variation,
ii. Study mating systems,
iii. Identify gene flow *via* both pollen and seeds,
iv. Reveal uniparental lineage and
v. Detect hybridization and introgression
vi. Analyze the genetic diversity and phylogeography in a wide range of plant species.

The major constraint of employing chloroplast SSR is that the prior requirement of chloroplast sequence data for primer designing. Though primers can be designed from the partially sequenced chloroplast genes or genomes and can be used to generate polymorphic PCR products from the species from which the sequences were obtained and their close relatives, the transportability to unrelated taxa, or even within the genus, is limited.

Despite this, universal primers to amplify chloroplast SSRs have been designed to amplify chloroplast microsatellite regions of dicotyledonous angiosperms (Weising and Gardner 1999). Especially, primers pairs that flanked A or T mononucleotide repeats ($n = 10$) of the tobacco chloroplast genome were found to be useful as potential genetic markers in Actinidiaceae, Brassicaceae, and Solanaceae group of plants. Similarly, common primer sets for the PCR amplification of chloroplast SSRs for all the plants that come under Poaceae (grass plants) have also been reported.

However, the development and employment of chloroplast-specific molecular markers have not yet been documented in Moringa and this laboratory has already made a preliminary analysis of publicly available chloroplast genome sequences for the development of new marker systems.

Generally, plant mitochondrial DNA (mtDNA) is larger in size, when compared with animal mtDNA. Usually, animal mtDNA has a size of 10 MDa per mitochondrial genome whereas plant mtDNA is 320 MDa and found to be complex (Sederoff et al. 1984). Further, relatively higher heterogeneity is reported in plant mtDNA, and it is normally characterized as circular chromosomes with varying size and abundance.

Because of the above reasons and their frequent reorganization of DNA sequences, mtDNA is not preferred for phylogenetic analysis. Despite this, mitochondrial haplotype diversity with respect to DNA sequence rearrangement has been explored in population differentiation and trait-based segregation of population in pine and fir taxa. Nevertheless, microsatellite DNA-based studies have not yet been reported in Moringa.

10.4.3.1 DNA Barcoding Genes

DNA barcoding has been launched with the central idea of providing a public resource of DNA sequences that can be used for the identification of biological units with their taxonomic classification. DNA barcoding was first productively established in animals by employing a small region of a mitochondrial gene, *cytochrome oxidase 1 (CO1).*

The vital characteristics of *CO1* that made it recognize it as a gold standard for animal DNA barcoding are universality, minimalism, and scalability. Generally, the selected DNA barcoding region should be amplified across the diverged and large animal samples and sequenced consistently and reliably. Besides they should enable effortless comparison of sequence data that held to differentiate the investigated species from one another.

Though *CO1* was found to be useful as a standard DNA barcoding gene, it was still challenging to find an appropriate gene in plants. As discussed in the last few lines of Sect. 10.4.3, a poor degree of nucleotide substitution in plant mitochondrial genomes, disqualify the employment of *CO1* as a widespread plant barcode. Owing to this, appropriate plant DNA barcoding regions have been searched outside the mitochondrial genome (i.e., in chloroplast and nuclear genome) and still there is no universal plant DNA barcode region has been discovered. Therefore, it is unanimously suggested that employing multiple DNA barcoding genes can be used to reveal plant species discrimination.

Accordingly, three different research groups/research consortia from the worldwide systematic community have initially recommended (based on in silico and laboratory-based assessments of the suitability of various markers) combinations of seven different plastid markers *viz.*, *rpoC1* + *rpoB* + *matK* or *rpoC1* + *matK* + *trnH-psbA* or *rbcL* + *trnH-psbA* and *atpF-H* + *psbK-I* + *matk* for plant DNA barcoding. The success of these combinations in plant barcoding has been discussed at the 2nd International Barcode of Life Conference in Taipei, and it was decided that such combinations are not as useful in the plant as that of the animal gold standard, *CO1*. Consequently, it was proposed that *matK*, *ITS*, and *rbcL* may be employed initially as plant DNA barcodes and if there is no convincing clarification/discrimination at the species level, it can proceed with other plant barcoding regions. Detailed implications of DNA barcoding in Moringa are provided in Chap. 9.

These barcodes were found to be easily amplified, sequenced, and aligned in the majority of land plants and shown to their utility in plant barcoding. As these markers were derived from the coding regions, in silico translation of the investigated sequences facilitates effortless verification of editing or assembly errors and the existence of pseudogenes. This process will also aid in correcting the sequence orientation which enables comparative analyses of diversity amongst taxonomic groups and geographical regions.

Sixty Moringa accessions collected from diverse geographical regions of South India are being maintained at Moringa Genetic Resource Garden at this institute, and their extent of variability was investigated based on morphological, biochemical, and molecular parameters. Comparatively better post-harvest quality parameters of leaves such as moisture loss percentage, color retention, and texture were found to be best in accession MO 8 even after 24 h of harvest. On the other hand, several other Moringa accessions recorded the highest score for yield and other post-harvest quality parameters of pods.

The results of DNA barcoding with *matk, rbcL,* and *ITS* regions helped to differentiate major types of Moringa accessions. We have completed DNA barcoding of several Moringa lines (NCBI GenBank IDs: KT737744 to KT737801), for the first time in the world, and more than 1500 chemically induced mutants under the background of Moringa cultivar PKM1 are being maintained at the Moringa Genetic Resources Garden along with other germplasm accessions that were collected from different geographical regions.

10.5 Future of Moringa Genome Sequencing/Resequencing Strategies

It is expected that all the workflow involved in sequencing starting from DNA or RNA isolation from a single cell and library preparation from little starting material will be simplified and automated in near future, which will generate longer reads with reduced error rates. This will facilitate to utilize the poor quality samples (such as noninvasive sampling from museums or fossils) to monitor historical genetic diversity. It is also anticipated within few years that novel tools and strategies will be developed to automate the assembly and annotation process of the huge amount of deep and lengthy sequencing data.

Since the expenses related to genome resequencing and bioinformatics analysis will be reduced due to rapid progress in genomics tools, it is possible to resequence all the available Moringa ecotypes and a great wealth of genomic information will be available to a larger spectrum of users. This implies that even resource-limited laboratories can explore the Moringa genomic resources for its genetic improvement. It is also imperative to design sound strategies and work plans for storing and sharing a huge amount of Moringa genomic and other omics data for its efficient utilization.

Availability of portable and fast sequencing of the targeted genome (such as MinION developed by Oxford Nanopore Technology) will enable its direct application in the field itself, and it will increase the efficiency of Moringa molecular breeding.

A bright future for genome mining is expected in the coming years to discover novel genes involved in the expression of nutritionally and medicinally useful products in Moringa and already several multinational biotechnology companies have started to work on this exciting research arena. They have employed genome mining approaches for Moringa natural product-based drug discovery and in order to obtain the proprietary rights, the details of these strategies have not been fully disclosed.

Thus, it is strongly believed that transition from Moringa genotype conservation genetics and genome-scale data to applied Moringa breeding will soon be realized, which will enable to ensure global economic, food, and nutritional security.

References

Boopathi NM (2020) Genetic mapping and marker assisted selection: basics, practice and benefits, 2nd edn. Springer Nature Singapore Pte Ltd., pp 516. https://doi.org/10.1007/978-981-15-2949-8_12

Chang Y, Liu H, Liu M, Liao X, Sahu SK, Fu Y, Song B, Cheng S, Kariba R, Muthemba S, Hendre PS (2019) The draft genomes of five agriculturally important African orphan crops. GigaScience 8(3):giy152

Iffiú-Soltész Z, Wanecq E, Lomba A, Portillo MP, Pellati F, Szökő É, Bour S, Woodley J, Milagro FI, Martinez JA, Valet P (2010) Chronic benzylamine administration in the drinking water improves glucose tolerance, reduces body weight gain and circulating cholesterol in high-fat diet-fed mice. Pharmacol Res 61(4):355–363

Laver T, Harrison J, O'neill PA, Moore K, Farbos A, Paszkiewicz K, Studholme DJ (2015) Assessing the performance of the oxford nanopore technologies minion. Biomol Detect Quantif 3:1–8

Lin W, Dai S, Chen Y, Zhou Y and Liu X (2019). The complete chloroplast genome sequence of *Moringa oleifera* Lam. (Moringaceae), Mitochondrial DNA Part B, 4(2):4094–4095

Liu J, Cai H, Li H, Liu Z, Zheng D, Shi C, Niu Y (2019) The chloroplast genome of *Moringa oleifera* (Moringaceae). Mitochondrial DNA Part B 4(1):646–647

Mlotshwa S, Pruss GJ, MacArthur JL, Endres MW, Davis C, Hofseth LJ, Peña MM, Vance V (2015) A novel chemopreventive strategy based on therapeutic microRNAs produced in plants. Cell Res 25(4):521–524

Ojeda-López J, Marczuk-Rojas JP, Polushkina OA, Purucker D, Salinas M, Carretero-Paulet L (2020) Evolutionary analysis of the Moringa oleifera genome reveals a recent burst of plastid to nucleus gene duplications. Sci Rep 10(1):1–15

Pasha SN, Shafi KM, Joshi AG, Meenakshi I, Harini K, Mahita J, Sajeevan RS, Karpe SD, Ghosh P, Nitish S, Gandhimathi A, Showdhamini R (2020) The transcriptome enables the identification of candidate genes behind medicinal value of Drumstick tree (Moringa oleifera). Genomics 112(1):621–628

Pirrò S, Zanella L, Kenzo M, Montesano C, Minutolo A, Potestà M, Sobze MS, Canini A, Cirilli M, Muleo R, Colizzi V (2016) MicroRNA from Moringa oleifera: identification by high throughput sequencing and their potential contribution to plant medicinal value. PLoS One 11(3):e0149495

Sederoff RR (1984) Structural variation in mitochondrial DNA. Adv Genet 22:1–108

Shu J, Chiang K, Zempleni J, Cui J (2015) Computational characterization of exogenous microRNAs that can be transferred into human circulation. PLoS One 10(11): e0140587

Tello-Ruiz MK, Naithani S, Gupta P, Olson A, Wei S, Preece J, Jiao Y, Wang B, Chougule K, Garg P, Elser J (2020) Gramene 2021: harnessing the power of comparative genomics and pathways for plant research. Nucl Acids Res

Tian Y, Zeng Y, Zhang J, Yang C, Yan L, Wang X, Shi C, Xie J, Dai T, Peng L, Huan YZ (2015) High quality reference genome of drumstick tree (Moringa oleifera Lam.), a potential perennial crop. Sci China Life Sci 58(7):627–638

Wang X, Shi Y, Chen Y, Fan J, Pu Y, Huang A (2020) Comparative proteome analysis of matured dry and germinating moringa oleifera seeds provides insights into protease activity during germination. Food Res Int 109332

Weising K, Gardner RC (1999) A set of conserved PCR primers for the analysis of simple sequence repeat polymorphisms in chloroplast genomes of dicotyledonous angiosperms. Genome 42(1):9–19

Xia E, Tong W, Hou Y, An Y, Chen L, Wu Q, Liu Y, Yu J, Li F, Li R, Li P (2020) The reference genome of tea plant and resequencing of 81 diverse accessions provide insights into its genome evolution and adaptation. Mol Plant 13(7):1013–1026

Xiao T, Zhou W (2020) The third generation sequencing: the advanced approach to genetic diseases. Transl Pediatr 9(2):163–173

Xu G, Ma H, Nei M, Kong H (2009) Evolution of F-box genes in plants: different modes of sequence divergence and their relationships with functional diversification. Proc Natl Acad Sci 106(3):835–840

Zhang L, Hou D, Chen X, Li D, Zhu L, Zhang Y, Li J, Bian Z, Liang X, Cai X, Yin Y (2012) Exogenous plant MIR168a specifically targets mammalian LDLRAP1: evidence of cross-kingdom regulation by microRNA. Cell Res 22(1):107–126

Moringa Functional Genomics: Implications of Long Read Sequencing Technologies

11

J. Deepa, Rohit Kambale, A. Bharathi, M. Williams, N. Manikanda Boopathi, and M. Raveendran

Abstract

The caloric needs of the global population are solely dependent on only 30 crops including rice, wheat, and maize. It is important to diversify and ensure the global food supply by enhancing the wide adaptation of nutritionally important vegetable crops. Focusing on the underutilized and locally available vegetable crops would meet the need of eradicating malnutrition and food security for the local population and small-hold farmers. With the advent of long-read sequencing technologies, several orphan crops and crops with complex genetic nature are being explored and genome resources are made available for crop improvement. *Moringa oliefera* is a well-known medicinal and nutritive plant which is getting adequate attention in recent years. Functional analytical tools including genome, transcriptome, metabolome, and proteome were made available for moringa; besides diversity analysis, evolutionary and syntenic studies were also reported. With an objective of exploring the available resources of an important nutritive crop and evaluating the need to enhance the functional studies further, this chapter reviews the application of long-read sequencing and their application in moringa genomics for its wider utilization in pharma industries.

11.1 Prelude on Functional Genomics Tools and Applications in Moringa

11.1.1 Tools and Strategies Available to Explore Moringa Functional Genomics

Promoting the research of orphan crops by generating genome and transcriptome sequences will provide insights into the genes involved in important agronomic traits (Jamnadass et al. 2020). Moringa (*Moringa oleifera*) is an important vegetable crop in many developing countries due to its medicinal and nutritional properties. Moringa is a softwood tree that belongs to the family *Moringaceae,* and it is originated from the sub-Himalayan region of India (Ramachandran et al. 1980). The Moringa plant is gaining attention in the recent past due to its nutritional, stress-tolerant, economical, and medicinal values. Each part of Moringa is reported to possess medicinal value including leaf extract, seed, stem, flower, and root (Anwar et al. 2007; Sehgal et al. 2012; Al-Asmari et al. 2015). Genome and

J. Deepa · R. Kambale · A. Bharathi · M. Williams · N. M. Boopathi · M. Raveendran (✉)
Department of Plant Biotechnology, Centre for Plant Molecular Biology & Biotechnology, Tamil Nadu Agricultural University, Coimbatore, India
e-mail: biotech@tnau.ac.in

N. M. Boopathi et al. (eds.), *The Moringa Genome*, Compendium of Plant Genomes,
https://doi.org/10.1007/978-3-030-80956-0_11

transcriptome sequencing of Moringa was decoded very recently (Tian et al. 2015; Chang et al. 2019; Panes et al. 2015). Such sequence information on medicinal plants will assist the understanding of biosynthesis pathways of medicinally important compounds.

Advances in sequencing technologies are evolving very rapidly from Sanger sequencing to long-read sequencing technologies. Next generation sequencing (NGS) platforms are mainly distinguished by their read length for instance. Second-generation sequencing technologies are capable of generating shorter read lengths (35–600 bp) compared to third-generation or long-read sequencing technologies (>1 kb). The major demerits of second-generation sequencing include data assembly and the inability to handle repeat sequences or large genomic rearrangements (Rhoads and Au 2015). The limitations of second-generation sequencing technologies can be overcome by long-read technologies as they can generate reads up to 2 Mb. Third-generation sequencing technologies including Pacbio, Nanopore, synthetic long reads, optical mapping, RNA-seq, hybrid sequencing (both second and third-generation techniques), and single-cell RNA/DNA sequencing are being widely adopted in the current decade. However, more advances in sequencing technologies are being evolved including advances in nanopore sequencing, in situ nucleic acid sequencing, microscopy-based sequencing (Kumar et al. 2019). The application of these advanced sequencing technologies has assisted the improvement of genome assemblies and transcriptome studies in many crops and also they served as a platform for analyzing underutilized crops. Such studies will promote the knowledge and understanding of the biochemical pathways of the important secondary metabolites and other chemical components in the medicinal crops resulting in dissecting the drug discovery mechanisms.

11.1.2 Third Generation Sequencing Technologies for Moringa Functional Genomics

Third-generation sequencing or long-read sequencing technologies hold numerous advantages compared to second-generation sequencing technologies. These long read technologies produce longer reads up to 10 kb compared to second-generation sequencing technologies which provide around ~600 bp read length (Amarasinghe et al. 2020). Moreover, third-generation sequencing technologies improve the efficiency of de novo assembly as second-generation sequencing technologies (such as HiSeq, MiSeq, NovaSeq, BGISEQ, and Ion torrent) involves challenges in genome assembly constitution with shorter reads (Dumschott et al. 2020).

Pacific Biosciences' (PacBio) single-molecule real-time (SMRT) sequencing and Oxford Nanopore Technologies' (ONT) nanopore sequencing are the two major long-read sequencing technologies widely being adopted for crop species sequencing in the current sequencing era. SMRT involves sequencing by synthesis whereas the ONT approach involves a novel technique in which the individual DNA molecule moves through the pore and sensors detect the changes in the ionic current according to the passing nucleotide, and this information will be sued for base calling (Deamer et al. 2016). So far, ONT has been applied in a wide range of crop plants from model crops to non-model crops, and it benefits genomes with highly repetitive regions. Notable species sequenced using the ONT platform are Arabidopsis (Michael et al. 2018), rice (Mondal et al. 2018; Read et al. 2020; Tanaka et al. 2020; Choi et al. 2020), teak (Yasodha et al. 2018), sorghum (Deschamps et al. 2018), yam (Siadjeu et al. 2020), brassica (Belser et al. 2018), and tomato (Schmidt et al. 2017).

The first reference genome for duckweed (*S. intermedia*) was developed using Pacbio, and ONT platform and genomic sequences were compared with its sister species *S. polyrhiza*. Both species revealed more than 20,000 putative protein-coding genes, very low rDNA copy numbers, and a low amount of repetitive sequences, mainly Ty3/gypsy retroelements. This study also detected few new small chromosome rearrangements between both Spirodela species which refined the karyotype and the chromosomal sequence assignment for *S. intermedia* (Hoang et al. 2020). One of the major limitations of these long-read technologies is the high error rate compared to second-generation sequencing technologies, and it requires high-quality DNA.

However, considering the advantages over previous generation sequencing technologies third-generation sequencing technologies will become an essential sequencing tool with its upcoming rapid advances in plant genomics especially for large and complex genomes (Dumschott et al. 2020). As outlined in Fig. 11.1, a large array of functional genomics tools is available to harness maximum biological information from the plants'cells and these tools are continuously evolving.

11.1.3 Pan Genome Studies for Functional Genomics

In the genome sequencing era, the concept of a single reference genome will not be efficient as the single genotype may not represent the complete genome information of a species. Pan-genome is a core genome that represents the total gene count of a species including copy number variations (CNVs), presence/absence variations (PAVs), and SNPs (Munir et al. 2020). Initial methods involved assembling the whole genomes of various individuals and comparing the sequence variation, whereas the current methods involve identification of PAVs and non-assembling reads of individuals are assembled and added to pan-genome; recent methods also includes graph-based assembly which describes conservation and diversity in a species

The ever-increasing sequencing projects for the important crop species demands the need for pan-genome studies. The first pan-genome sequence was demonstrated in bacteria (Tettelin et al. 2005). In recent times, pan-genome studies play a major role in functional genomics. The first plant pan-genome assembly was made for soybean (Li et al. 2014). This study involved

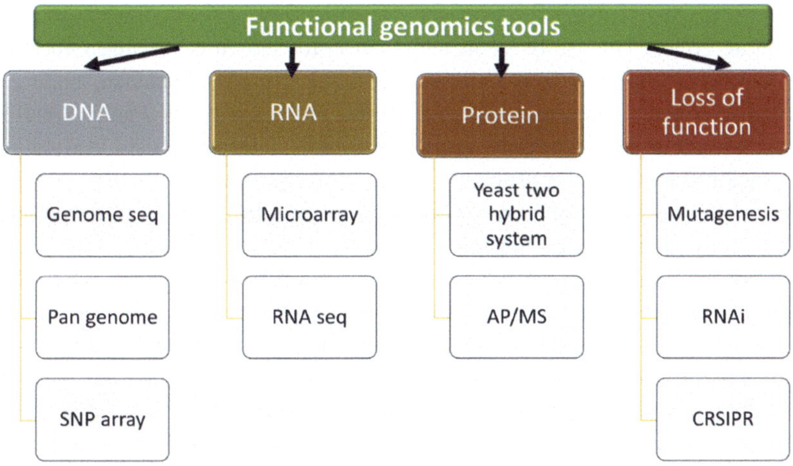

Fig. 11.1 Various functional analysis tools. (1) DNA level analysis which includes genome sequencing and structural analysis (2) RNA level analysis which includes transcriptome analysis and gene expression studies (3) Protein level analysis which includes proteomic studies revealing an entire set of protein in a species and their role (4) Mutation studies which include the creation of artificial mutation to study the function of a gene including Mutagenesis and CRISPR gene editing for functional analysis

core genome analysis of seven wild individuals resulting in the identification of genes with copy number variation, large-effect mutations, and positive selection relating to variations in agronomic traits including biotic stress, seed composition, flowering and maturity, organ size, and biomass.

Pan-genomes are being made available for many important crop species including Arabidopsis (Gan et al. 2011), rice (Schatz et al. 2014; Yao et al. 2015; Zhao et al. 2018; Wang et al. 2018), Brassica (Lin et al. 2014), soybean (Li et al. 2014; Liu et al. 2020), maize (Hirsch et al. 2014), wheat (Montenegro et al. 2017), capsicum (Ou et al. 2018), tomato (Gao et al. 2019), and walnut (Trouern-Trend et al. 2020).

Zhao et al. (2018) developed a pan-genome of 66 rice accessions belonging to *Oryza rufipogon* and *Oryza sativa*. This study identified 23 million sequence variations in the rice genome. They also reported the functional variants of five important quantitative trait loci (QTLs)—Hd3a (Os06g0157700), COLD1 (Os04g0600800), GW6a (Os06g0650300), TAC1 (Os09g0529300), and Sd1 (Os01g0883800), which are involved in flowering time, cold tolerance, grain weight, tiller angle, and plant height, respectively. A recent study in soybean reported the assembly of graph-based single reference genome constructed using 26 soybean accessions (Liu et al. 2020). This study identified large structural variations and gene fusion events and their functional role in gene expressions and agronomic traits. Pan-genome of tomato comprising 725 accessions revealed the absence of 4,873 genes from the reference genome (Gao et al. 2019). This study also revealed the role of *TomLoxC* promotor in controlling fruit flavor. Such pan-genome studies in important crop plants are becoming the source of variations and provide insights into domestication which will greatly assist breeding for better agronomic traits. However, pan-genome studies in Moringa have not yet been reported.

11.1.4 Long Read Sequencing Technologies for Transcriptomic Studies

Illumina's short read sequencing platform, RNA-Seq has been popularly used for plant transcriptomic studies in the last decade. As short-read sequencing involves fragmentation of the full-length cDNA, loss of information from the original transcripts becomes unavoidable, and it is one of the major drawbacks of short-read sequencing platforms (Cui et al. 2020). Long read-sequencing platforms such as PacBio, ONT can sequence the full-length cDNA thus overcoming the loss of information which helps in analyzing the post-transcriptional events. The long-read sequencing platforms have been widely adopted in animals and being applied in plant transcriptomic studies especially since many uncharacterized species are being explored. In order to apply both sequencing platforms and to compare the results of these sequencing technologies, many studies have applied both Illumina and PacBio and /or ONT sequencing platforms for transcriptomic studies.

Several economically valued plants such as ornamental crops are being sequenced, and the understanding of the important characteristics such as flower color, fragrance are being carried out in the recent past using long-read sequencing technologies. For instance, Huang et al. (2020) utilized Illumina and PacBio platforms to study the molecular mechanism leading to the variation in flower color in ornamental crabapple. This study identified 603 differentially expressed genes (DEGs), including 449 upregulated and 154 downregulated genes. The role of transcription factors related to anthocyanin synthesis was reported, and five genes related to anthocyanin transport and degradation were found to be highly expressed in red petals. *R. lapponicum* is an important ornamental crop that possesses high economic value worldwide. Recently, Jia et al.

(2020) reported a full-length transcriptome of *R. lapponicum*. Based on KEGG analysis, this study identified 96 transcripts coding for the enzymes associated with anthocyanin synthesis.

Similarly, Zhang et al. (2019a) employed both Illumina and PacBio platforms in order to study the transcripts of brown planthopper (BPH) which is one of the major pests of rice. This study revealed the full diversity and complexity of the BPH transcriptome and indicates that BPH responses to rice resistance might be related to starvation stress responses, nutrient transformation, oxidative decomposition, and detoxification. PacBio sequencing has also been applied for transcriptomic studies of other important food crops including wheat (Dong et al. 2015), maize (Zhou et al. 2018; Wang et al. 2016), sorghum (Abdel-Ghany et al. 2016), coffee (Cheng et al. 2017), and garlic (Chen et al. 2018).

11.2 Advances in Moringa Functional Genomics

The high nutritional content of *M.oleifera* such as high protein, vitamin, and mineral content makes it attractive and widespread in developing countries. Every part of *M.oleifera* possess medicinal values, and it has been extensively studied for its nutritional and medicinal importance in treating diseases. For instance, anti-inflammatory, antimicrobial properties of root, leaf, stem, flower, and pod are reported to lower back pain, antidiabetic, and anti-cancer (Goyal et al. 2007). It is highly adapted to arid and semi-arid tropics due to its drought tolerance.

Genome sequencing and transcriptome sequencing of medicinal crops are being made available, and the pharmacological impacts are analyzed in detail for drug discovery. With the advent of NGS technologies, moringa draft genome and transcriptome are made available to explore the possibilities of understanding and enhancing its stress tolerance, nutritional, and pharmaceutical mechanisms.

The first transcriptome study on Moringa pods was demonstrated by Panes et al. (2015). This study generated a total of 182,588 transcripts out of which 3,556 unigenes were found to be involved in oil biosynthesis. Moreover, most of the unigenes were found to be involved in fatty acid biosynthesis with 1,009 unigenes, fatty acid catabolism with 982 unigenes, and triacylglycerol catabolism with 608 unigenes. Around 33 unigenes encodings for transcription factors were reported to be involved in regulating oil biosynthesis gene expression. This transcriptome resource for the *M. oleifera* Lam. mature seed embryo would assist in mapping of oil biosynthesis-related genes and the understanding of metabolic pathways which could possibly be used to improve seed yield and oil content of *M. oleifera*.

Draft genome of *M. oleifera* was reported by Chang et al. (2019) along with four other agriculturally important plants. This study predicted 18,451 protein-coding genes. Gene expansion and contraction study assisted the characterization of root nodule symbiosis genes, transcription factors, and starch biosynthesis-related genes in the five genomes. Moringa seeds are can be a good source of edible oil as the seeds are capable of producing oil up to 40%.

Pasha et al. (2020) reported the transcriptome of leaf, root, stem, seed, and flower of moringa cultivar, Bhagya. This study predicted 17,148 gene models and candidate genes related to the biosynthesis of secondary metabolites, vitamins, and ion transporters were identified. They also performed expression analysis through RT-PCR and metabolite quantification which showed a high expression pattern in the leaves, flowers, and seeds of the genes/enzymes involved in the biosynthesis of vitamins and metabolites like quercetin and kaempferol. In addition, this study revealed the expression of iron transporters and calcium storage proteins were observed in root and leaves and concluded that leaves retain the highest number of small molecules of interest. In continuation, assessment of the combined transcriptome for transcript abundance across five tissues assisted the prediction of the protein-coding genes. Further, these identified protein-coding genes from the transcripts were annotated and used for orthology analysis (Shafi et al. 2020).

Gene family evolution in Moringa was reported very recently by Lopez et al. (2020). This study reported gene expansion of 101 gene families grouping 957 genes, and the expanded families were highly enriched for chloroplastic and photosynthetic functions. In addition, this study also reported the large regions of plastid DNA (4.71%) insertions into the nuclear genome through microsynteny analysis of ten other plant species including rice, maize, and Arabidopsis.

11.2.1 Characteristics of Moringa Plastid Genome

The complete chloroplast genome of *M. oleifera* was reported by Liu et al. (2019a). The reported chloroplast genome size is 160,599 bp long and includes 113 full-length genes including 79 protein-coding genes. Its large single copy (LSC), small single copy (SSC), and inverted repeat (IR) regions are 88,933, 19,482, and 26,092 bp long, respectively. Phylogenetic tree analysis exhibited that *M. oleifera* was clustered with other Moringaceae species with 100% bootstrap values.

Another study by Mu et al. (2019) reported plastid genome sequence for the four species belonging to brassicales family including *M. oleifera*. In comparison to Liu et al. (2019a), the size of the plastid genome of *M. oleifera* was 163,131 bp and possess typical quadripartite structure: IRs, LSC, and SSC. The length of LSC, IR, and SSC regions of *M. oleifera* Lam. are 102,342 bp; 3,710 bp; 48,715 bp, respectively. All four species had the same number of 78 protein-coding genes, four ribosomal RNAs, while the number of transfer RNAs varies from 36 (*Cleome rutidosperma* DC.), 37 (*Carica papaya* L. and *Moringa oleifera* Lam.) to 38 (*Capparis urophylla* F.Chun).

More recently, Lin et al. (2019b) reported the complete chloroplast genome which was very close to the previously reported chloroplast genomes by Liu et al. (2019a) and Mu et al. (2019). The chloroplast genome was 160,600 bp in length with 88,577 bp of LSC, 18,883 bp of SSC, and 26,570 bp of IR. This study predicted 131 genes, and the phylogenetic analysis of 71 protein-coding sequences of 13 plant plastomes showed that the *M. oleifera* is closest to *Carica papaya*.

11.2.2 Functional Studies in Moringa

Availability of genome sequencing data provides opportunities to explore functional studies such as synteny, evolutionary and phylogenetic analysis using bioinformatics tools. Deng et al. (2016) screened 18 candidate genes selected from the Moringa transcriptome database. Expression stabilities of the selected agronomically important traits were examined in 90 samples collected from the pods in different developmental stages, various tissues, and the roots and leaves under different conditions (low or high temperature, sodium chloride (NaCl)- or polyethyleneglycol (PEG)- simulated water stress).

This study provided insights on the Moringa genes involved in abiotic stress tolerance and forms a basis for Moringa functional gene analysis. The first proteomic analysis of the flower of *M. oleifera* was reported by Shi et al. (2018). This study identified 9443 peptides corresponding to 4004 high-confidence proteins and a number of commercially important food-grade enzymes were also commented. A total of 261 proteins were annotated as carbohydrate-active enzymes, 16 proteases, 22 proteins are assigned to the citrate cycle, which the top proteins were assigned to the GH family, cysteine synthase, and serine/threonine-protein phosphatase. These enzymes indicated that they are a new source with potential use for fermentation and brewing industry, fruit and vegetable storage, and the development of functional peptides.

WRKY transcription factors are known to be involved in numerous plant processes from germination to senescence. With the resource available on the Moringa genome (Tian et al. 2015), genome-wide identification and characterization of WRKY transcription factors were reported by Zhang et al. (2019b). This study identified 54 MoWRKY TFs, and the expression analysis using RT-PCR revealed the involvement of potential MoWRKY genes with respect to abiotic stresses such as salt, heat, drought, H_2O_2, and cold.

A recent study reported the genetic diversity of 57 M. *oleifera* accessions using RAPD marker and their biologically active component such as cinnamic, caffeic, ferulic, and coumaric acids, flavonoids analysis using HPLC (Panwar and Mathur 2020). This study grouped the 57 accessions into five groups, and high diversity in the concentration of active compounds was also reported using HPLC. The strong correlation between phytochemical variables and DNA polymorphism will assist in breeding for selecting the best accessions.

11.2.3 Metabolomics in Moringa

As moringa possess versatile utility for medicine and nutrition sources, profiling of its metabolites and making the common metabolite database of various tissues are very important for functional studies. Mahmud et al. (2014) reported the profiling of metabolites from moringa leaf and stem tissues. Among the 30 metabolites identified in this study, 22 metabolites were common in both leaf and stem tissues and the remaining eight metabolites included, 4-aminobutyrate, adenosine, guanosine, tyrosine, and p-cresol were found only in leaf tissues; whereas, glutamate, glutamine, and tryptophan were found only in stem tissues. They also performed biochemical pathway analysis which revealed that 28 identified metabolites were interconnected with 36 different pathways as well as related to different fatty acids and secondary metabolites synthesis biochemical pathways.

Flavonoids are important secondary metabolites with specific metabolic functions in plants. *Moringa oleifera* and *M. ovalifolia* are two moringa species known to contain a wide spectrum of flavonoids molecules with known nutraceutical properties. A comparative analysis of flavonoid content in *M. oleifera* and *M. ovalifolia* with the aid of ultra-high-performance liquid chromatography coupled with high-resolution quadrupole time-of-flight mass spectrometer (UHPLC-qTOF-MS) fingerprinting was demonstrated by Makita et al. (2016). Various flavonoids identified from these two species conclude that the various genetic bases of flavonoid biosynthesis in these species. The differentiation of the flavonoids among these species was mainly due to the superior glycosylation capabilities of *M. olifera* compared *M. ovalifolia*. This study concluded that *M. olifera* has wide pharmacological application based on its glycosylation complexity.

The metabolite and protein content of the plant are highly influenced by soil types. A recent study performed metabolite profiling of moringa leaves cultivated with vermicompost and phosphate rock under water stress conditions (Albores et al. 2019). UPLC-ESI–MS/MS analysis of leaf extracts revealed that the most abundant metabolites were flavonoids, alkaloids, and terpenes. This study identified that the water stress-induced changes in the metabolomics profile and the morphometric variables of *M. oleifera*.

A comparative study of the chemical constituents from moringa leaves collected from different cultivation regions, i.e., India and China were reported using liquid chromatography and mass spectroscopy (Lin et al. 2019c). This study reported a total of 122 characterized components, containing 118 shared constituents, from moringa leaves of India and China. Such a comprehensive phytochemical profile study provides insights into the basis for explaining the effect of different growth environments on secondary metabolites.

A very recent study by Rocchetti et al. (2020) had comprehensively investigated the (poly)-phenolic profile of *M. oleifera* leaves through untargeted metabolomics, following a homogenizer-assisted extraction (HAE) using three solvent systems, i.e., methanol (HAE-1), methanol–water 50:50 v/v (HAE-2), and ethyl acetate (HAE-3). They annotated close to 300 compounds, recording mainly flavonoids and phenolic acids. In addition, they also reported antioxidant capacity, antimicrobial activity, and enzyme inhibition assays in the different extracts.

This study concluded that *M. oleifera* leaf extracts are a good source of bioactive polyphenols with potential use in the food and pharma industries.

11.3 Medicinal and Pharmaceutical Studies on *M. oleifera*

Several studies have reported the benefits of the Moringa plant (leaf, flower, seed, stem) and its extracts in controlling non-communicable diseases such as diabetes, obesity, cancer, heart disease, and stroke (Lin et al. 2018). Though a smaller number of studies are available on humans, many studies on animals have reported a positive association between the consumption of *M. oleifera*-containing foods and a reduced risk of developing certain types of NCDs. For instance, Li et al. (2020) reported transcriptome gene expression and epigenome DNA methylation in mouse kidney mesangial cells (MES13) using next-generation sequencing technology. After high glucose treatment, epigenome and transcriptome were found to be significantly altered and exposure to Moringa isothiocyanate (MIC-1) which is a bioactive constituent found abundantly in *M. oleifera*, possesses antioxidant and anti-inflammation properties reversed some of the changes caused by high glucose.

Another study by Cheng et al. (2019) reported the *Nrf2-ARE* antioxidant activity of MIC-1, and its potential in diabetic nephropathy. In brief, this study concluded that MIC-1 activates *Nrf2-ARE* signaling, increases expression of *Nrf2* target genes, and suppresses inflammation, while also reducing oxidative stress and possibly *TGFβ1* signaling in high glucose-induced renal cells. Sun et al. (2019) showed the positive effects of Moringa leaf extract on the treatment of type 2 diabetes mellitus by influencing the expression *Per1* and *Per2* genes. This study concluded that moringa leaf extract can ameliorate liver damage in diabetic rats, possibly due to its anti-glycation activities. In addition, anti-inflammatory, anti-oxidative, and anti-cancer properties of MIC-1 were reported by Wang et al. (2019).

Natural plant-derived biostimulants are proven to improve the growth, yield, and post-harvest quality of horticultural products. Moringa leaf extracts in particular have been shown to improve seed germination, plant growth and yield, nutrient use efficiency, crop, and product quality traits (pre- and post-harvest), as well as tolerance to abiotic stresses (Zulfiqar et al. 2020). The use of plant-derived biostimulants such as moringa leaf extracts can help in reducing the fertilizer quantities needed and thus contribute to achieving global food security sustainably.

11.4 Future perspectives

Though the medicinally important moringa crop is being explored in the recent past, there is a need to improve its genomics resources such as markers, QTLs, and candidate gene identification for crop improvement. It is important to explore genome-wide analysis with an increased number of accessions for diversity studies and understanding potential marker-trait association. Such studies with available third-generation sequencing technologies will provide millions of SNPs leading to the development of SNP-chip arrays which can serve as rich marker resources especially for marker-assisted selection. Genome sequence of various accessions of moringa will pave way for developing pan-genomes which will assist the identification of sequence-level variation such as CNVs and PAVs. CNVs have potential effects on gene expression and structure relating to phenotypic changes in various accessions. Developing the core genome of moringa will identify the potential candidate genes involving in its abiotic stress-tolerant mechanism and biochemical pathways of important secondary metabolites contributing to its anti-inflammatory properties.

References

Abdel-Ghany SE, Hamilton M, Jacobi JL, Ngam P, Devitt N, Schilkey F, Ben-Hur A, Reddy AS (2016) A survey of the sorghum transcriptome using single-molecule long reads. Nat Commun 7(1):1–11

Al-Asmari AK, Albalawi SM, Athar MT, Khan AQ, Al-Shahrani H, Islam M (2015) Moringa oleifera as an anti-cancer agent against breast and colorectal cancer cell lines. PloS One 10(8):e0135814

Amarasinghe SL, Su S, Dong X, Zappia L, Ritchie ME, Gouil Q (2020) Opportunities and challenges in long-read sequencing data analysis. Genome Biol 21(1):1–16

Anwar F, Latif S, Ashraf M, Gilani AH (2007) Moringa oleifera: a food plant with multiple medicinal uses. Phytother Res Int J Devoted Pharmacol Toxicol Eval Nat Prod Deriv 21(1):17–25

Belser C, Istace B, Denis E, Dubarry M, Baurens FC, Falentin C, Genete M, Berrabah W, Chèvre AM, Delourme R, Deniot G (2018) Chromosome-scale assemblies of plant genomes using nanopore long reads and optical maps. Nat Plants 4(11):879–887

Chang Y, Liu H, Liu M, Liao X, Sahu SK, Fu Y, Song B, Cheng S, Kariba R, Muthemba S, Hendre PS (2019) The draft genomes of five agriculturally important African orphan crops. GigaScience 8(3):giy152

Chen X, Liu X, Zhu S, Tang S, Mei S, Chen J, Li S, Liu M, Gu Y, Dai Q, Liu T (2018) Transcriptome-referenced association study of clove shape traits in garlic DNA Res 25(6):587–596

Cheng B, Furtado A, Henry RJ (2017) Long-read sequencing of the coffee bean transcriptome reveals the diversity of full-length transcripts. Gigascience, 6 (11):gix086

Cheng D, Gao L, Su S, Sargsyan D, Wu R, Raskin I, Kong AN (2019) Moringa isothiocyanate activates Nrf2: potential role in diabetic nephropathy. AAPS J 21(2):31

Choi JY, Lye ZN, Groen SC, Dai X, Rughani P, Zaaijer S, Harrington ED, Juul S, Purugganan MD (2020) Nanopore sequencing-based genome assembly and evolutionary genomics of circum-basmati rice. Genome Biol 21(1):21

Cui J, Lu Z, Xu G, Wang Y, Jin B (2020) Analysis and comprehensive comparison of PacBio and nanopore-based RNA sequencing of the Arabidopsis transcriptome. Plant Methods 16(1):1–13

Deamer D, Akeson M, Branton D (2016) Three decades of nanopore sequencing. Nat Biotechnol 34(5):518–524

Deng LT, Wu YL, Li JC, OuYang KX, Ding MM, Zhang JJ, Li SQ, Lin MF, Chen HB, Hu XS, Chen XY (2016). Screening reliable reference genes for RT-qPCR analysis of gene expression in Moringa oleifera. PloS One 11(8):e0159458

Deschamps S, Zhang Y, Llaca V, Ye L, Sanyal A, King M, May G, Lin H (2018) A chromosome-scale assembly of the sorghum genome using nanopore sequencing and optical mapping. Nat Commun 9 (1):1–10

Dong L, Liu H, Zhang J, Yang S, Kong G, Chu JS, Chen N, Wang D (2015) Single-molecule real-time transcript sequencing facilitates common wheat genome annotation and grain transcriptome research. BMC Genomics 16(1):1039

Dumschott K, Schmidt MHW, Chawla HS, Snowdon R, Usadel B (2020) Oxford nanopore sequencing: new opportunities for plant genomics? J Exp Bot

Gan X, Stegle O, Behr J, Steffen JG, Drewe P, Hildebrand KL, Lyngsoe R, Schultheiss SJ, Osborne EJ, Sreedharan VT, Kahles A (2011) Multiple reference genomes and transcriptomes for Arabidopsis thaliana. Nature 477(7365):419–423

Gao L, Gonda I, Sun H, Ma Q, Bao K, Tieman DM, Burzynski-Chang EA, Fish TL, Stromberg KA, Sacks GL, Thannhauser TW (2019) The tomato pan-genome uncovers new genes and a rare allele regulating fruit flavor. Nat Genet 51(6):1044–1051

Goyal BR, Agrawal BB, Goyal RK, Mehta AA (2007) Phyto-pharmacology of Moringa oleifera Lam.—an overview. Nat Prod Radiance 6:347–353

Guzmán-Albores JM, Ramírez-Merchant ML, Interiano-Santos EC, Manzano-Gómez LA, Castañón-González JH, Winkler R, Abud-Archila M, Montes-Molina JA, Gutiérrez-Miceli FA, Ruíz-Valdiviezo VM (2019) Metabolomic and proteomic analysis of Moringa oleifera cultivated with vermicompost and phosphate rock under water stress conditions. Int J Agric Biol 21 (4):786–794

Hirsch CN, Foerster JM, Johnson JM, Sekhon RS, Muttoni G, Vaillancourt B, Peñagaricano F, Lindquist E, Pedraza MA, Barry K, de Leon N (2014) Insights into the maize pan-genome and pan-transcriptome. Plant Cell 26(1):121–135

Hoang PT, Fiebig A, Novák P, Macas J, Cao HX, Stepanenko A, Chen G, Borisjuk N, Scholz U, Schubert I (2020) Chromosome-scale genome assembly for the duckweed Spirodela intermedia, integrating cytogenetic maps, PacBio and Oxford Nanopore libraries. Sci Rep 10(1):1–14

Huang B, Rong H, Ye Y, Ni Z, Xu M, Zhang W, Xu LA (2020) Transcriptomic analysis of flower color variation in the ornamental crabapple (Malus spp.) half-sib family through Illumina and PacBio Sequel sequencing. Plant Physiol Biochem 149:27–35

Jamnadass R, Mumm RH, Hale I, Hendre P, Muchugi A, Dawson IK, Powell W, Graudal L, Yana-Shapiro H, Simons AJ, Van Deynze A (2020) Enhancing African orphan crops with genomics. Nat Genet 52(4):356–360

Jia X, Tang L, Mei X, Liu H, Luo H, Deng Y, Su J (2020) Single-molecule long-read sequencing of the full-length transcriptome of Rhododendron lapponicum L. Sci Rep 10(1):1–11

Kumar KR, Cowley MJ, Davis RL (2019) Next-generation sequencing and emerging technologies. In: Seminars in thrombosis and hemostasis, vol 45, no 07. Thieme Medical Publishers, pp 661–673. (2019, October)

Li S, Li W, Wu R, Yin R, Sargsyan D, Raskin I, Kong AN (2020) Epigenome and transcriptome study of moringa isothiocyanate in mouse kidney mesangial cells induced by high glucose, a potential model for diabetic-induced nephropathy. AAPS J 22(1):8

Li YH, Zhou G, Ma J, Jiang W, Jin LG, Zhang Z, Guo Y, Zhang J, Sui Y, Zheng L, Zhang SS (2014) De novo assembly of soybean wild relatives for pan-genome analysis of diversity and agronomic traits. Nat Biotechnol 32(10):1045–1052

Lin K, Zhang N, Severing EI, Nijveen H, Cheng F, Visser RG, Wang X, de Ridder D, Bonnema G (2014) Beyond genomic variation-comparison and functional annotation of three Brassica rapagenomes: a turnip, a rapid cycling and a Chinese cabbage. BMC Genomics 15(1):250

Lin M, Zhang J, Chen X (2018) Bioactive flavonoids in Moringa oleifera and their health-promoting properties. J Funct Foods 47:469–479

Liu J, Cai HH, Li HQ, Liu ZY, Zheng C, Shi C, Niu YF (2019a) The chloroplast genome of Moringa oleifera (Moringaceae). Mitochondrial DNA Part B 4(1):646–647

Lin W, Dai S, Chen Y, Zhou Y, Liu X (2019b). The complete chloroplast genome sequence of Moringa oleifera Lam.(Moringaceae). Mitochondrial DNA Part B, 4(2):4094–4095

Lin H, Zhu H, Tan J, Wang H, Wang Z, Li P, Zhao C, Liu J (2019c) Comparative analysis of chemical constituents of Moringa oleifera leaves from China and India by ultra-performance liquid chromatography coupled with quadrupole-time-of-flight mass spectrometry. Molecules 24(5):942

Liu Y, Du H, Li P, Shen Y, Peng H, Liu S, Zhou GA, Zhang H, Liu Z, Shi M, Huang X (2020) Pan-genome of wild and cultivated soybeans. Cell 182(1):162–176

Mahmud I, Chowdhury K, Boroujerdi A (2014) Tissue-specific metabolic profile study of Moringa oleifera L. using nuclear magnetic resonance spectroscopy.In: Plant tissue culture & biotechnology. Bangladesh Association for Plant Tissue Culture & Biotechnology; BAPTC&B, vol 24, no 1, pp 77

Makita C, Chimuka L, Steenkamp P, Cukrowska E, Madala E (2016) Comparative analyses of flavonoid content in Moringa oleifera and Moringa ovalifolia with the aid of UHPLC-qTOF-MS fingerprinting. S Afr J Bot 105:116–122

Michael TP, Jupe F, Bemm F, Motley ST, Sandoval JP, Lanz C, Loudet O, Weigel D, Ecker JR (2018) High contiguity Arabidopsis thaliana genome assembly with a single nanopore flow cell. Nat Commun 9 (1):1–8

Mondal TK, Rawal HC, Chowrasia S, Varshney D, Panda AK, Mazumdar A, Kaur H, Gaikwad K, Sharma TR, Singh NK (2018) Draft genome sequence of first monocot-halophytic species Oryza coarctata reveals stress-specific genes. Sci Rep 8(1):1–13

Montenegro JD, Golicz AA, Bayer PE, Hurgobin B, Lee H, Chan CK, Visendi P, Lai K, Doležel J, Batley J, Edwards D (2017) The pangenome of hexaploid bread wheat. Plant J 90(5):1007–1013

Mu W, Yang T, Liu X (2019) The complete plastid genomes of four species from Brassicales. Mitochondrial DNA Part B 4(1):124–125

Munir F, Saba NU, Arveen M, Siddiqa A, Ahmad J, Amir R (2020) Pan-genomics of plants and its applications. In: Pan-genomics: applications, challenges, and future prospects. Academic Press, pp 285–306

Ojeda-López J, Marczuk-Rojas JP, Polushkina OA, Purucker D, Salinas M, Carretero-Paulet L (2020) Evolutionary analysis of the Moringa oleifera genome reveals a recent burst of plastid to nucleus gene duplications. Sci Rep 10(1):1–15

Ou L, Li D, Lv J, Chen W, Zhang Z, Li X, Yang B, Zhou S, Yang S, Li W, Gao H (2018) Pan-genome of cultivated pepper (Capsicum) and its use in gene presence–absence variation analyses. New Phytol 220 (2):360–363

Panes VA, Kitazumi A, Butler M, Baoas R, De los Reyes BG (2015) Analysis of the oil biosynthesis transcripts of the Moringa oleifera Lam. mature seed embryos using RNA sequencing. In:I international symposium on Moringa, vol 1158, pp 55–62 (2015, November)

Panwar A, Mathur J (2020) Genetic and biochemical variability among Moringa oleifera Lam. accessions collected from different agro-ecological zones. Genome 63(3):169–177

Pasha SN, Shafi KM, Joshi AG, Meenakshi I, Harini K, Mahita J, Sajeevan RS, Karpe SD, Ghosh P, Nitish S, Gandhimathi A (2020) The transcriptome enables the identification of candidate genes behind medicinal value of Drumstick tree (Moringa oleifera). Genomics 112(1):621–628

Ramachandran C, Peter KV, Gopalakrishnan PK (1980) Drumstick (Moringa oleifera): a multipurpose Indian vegetable. Econ Bot 276–283

Read AC, Moscou MJ, Zimin AV, Pertea G, Meyer RS, Purugganan MD, Leach JE, Triplett LR, Salzberg SL, Bogdanove AJ (2020) Genome assembly and characterization of a complex zfBED-NLR gene-containing disease resistance locus in Carolina Gold Select rice with Nanopore sequencing. *PLoS genetics*, *16*(1), e1008571.

Rhoads A, Au KF (2015) PacBio sequencing and its applications. Genomics Proteomics Bioinform 13 (5):278–289

Rocchetti G, Pagnossa JP, Blasi F, Cossignani L, Piccoli RH, Zengin G, Montesano D, Cocconcelli PS, Lucini L (2020) Phenolic profiling and in vitro bioactivity of Moringa oleifera leaves as affected by different extraction solvents. Food Res Int 127:108712

Schatz MC, Maron LG, Stein JC, Wences AH, Gurtowski J, Biggers E, Lee H, Kramer M, Antoniou E, Ghiban E, Wright MH (2014) Whole genome de novo assemblies of three divergent strains of rice, Oryza sativa, document novel gene space of aus and indica. Genome Biol 15(11):506

Schmidt MH, Vogel A, Denton AK, Istace B, Wormit A, van de Geest H, Bolger ME, Alseekh S, Maß J, Pfaff C, Schurr U (2017) De novo assembly of a new Solanum pennellii accession using nanopore sequencing. Plant Cell 29(10):2336–2348

Sehgal N, Gupta A, Valli RK, Joshi SD, Mills JT, Hamel E, Khanna P, Jain SC, Thakur SS, Ravindranath V (2012) Withania somnifera reverses Alzheimer's disease pathology by enhancing low-density lipoprotein receptor-related protein in liver. Proc Natl Acad Sci 109(9):3510–3515

Shafi KM, Joshi AG, Meenakshi I, Pasha SN, Harini K, Mahita J, Sajeevan RS, Karpe SD, Ghosh P, Nitish S, Gandhimathi A (2020) Dataset for the combined transcriptome assembly of M. oleifera and functional annotation. Data in Brief, 105416.

Shi Y, Wang X, Huang A (2018) Proteomic analysis and food-grade enzymes of Moringa oleifer Lam. a Lam. flower. Int J Biol Macromol 115:883–890

Siadjeu C, Pucker B, Viehöver P, Albach DC, Weisshaar B (2020) High contiguity de novo genome sequence assembly of Trifoliate yam (Dioscorea dumetorum) using long read sequencing. Genes 11 (3):274

Sun W, Liu J, Wu L, Guo X, Zhang L, Fan Y, Yang L, Guo X, Hou Y, Mu X, Qin L (2019) Transcriptome analysis of the effects of Moringa Oleifera leaf extract in db/db mice with type 2 diabetes mellitus. Int J Clin Exp Med 12(6):6643–6658

Tanaka T, Nishijima R, Teramoto S, Kitomi Y, Hayashi T, Uga Y, Kawakatsu T (2020) De novo genome assembly of the indica rice variety IR64 using linked-read sequencing and nanopore sequencing. G3: Genes, Genomes, Genetics 10(5):1495–1501

Tettelin H, Masignani V, Cieslewicz MJ, Donati C, Medini D, Ward NL, Angiuoli SV, Crabtree J, Jones AL, Durkin AS, DeBoy RT (2005) Genome analysis of multiple pathogenic isolates of Streptococcus agalactiae: implications for the microbial "pan-genome." Proc Natl Acad Sci 102(39):13950–13955

Tian Y, Zeng Y, Zhang J, Yang C, Yan L, Wang X, Shi C, Xie J, Dai T, Peng L, Huan YZ (2015) High quality reference genome of drumstick tree (Moringa oleifera Lam.), a potential perennial crop. Sci China Life Sci 58(7):627–638

Trouern-Trend AJ, Falk T, Zaman S, Caballero M, Neale DB, Langley CH, Dandekar AM, Stevens KA, Wegrzyn JL (2020) Comparative genomics of six Juglans species reveals disease-associated gene family contractions. Plant J 102(2):410–423

Wang B, Tseng E, Regulski M, Clark TA, Hon T, Jiao Y, Lu Z, Olson A, Stein JC, Ware D (2016) Unveiling the complexity of the maize transcriptome by single-molecule long-read sequencing. Nat Commun 7(1):1–13

Wang C, Wu R, Sargsyan D, Zheng M, Li S, Yin R, Su S, Raskin I, Kong AN (2019) CpG methyl-seq and RNA-seq epigenomic and transcriptomic studies on the preventive effects of Moringa isothiocyanate in mouse epidermal JB6 cells induced by the tumor promoter TPA. J Nutr Biochem 68:69–78

Wang W, Mauleon R, Hu Z, Chebotarov D, Tai S, Wu Z, Li M, Zheng T, Fuentes RR, Zhang F, Mansueto L (2018) Genomic variation in 3,010 diverse accessions of Asian cultivated rice. Nature 557(7703):43–49

Yao W, Li G, Zhao H, Wang G, Lian X, Xie W (2015) Exploring the rice dispensable genome using a metagenome-like assembly strategy. Genome Biol 16 (1):1–20

Yasodha R, Vasudeva R, Balakrishnan S, Sakthi AR, Abel N, Binai N, Rajashekar B, Bachpai VK, Pillai C, Dev SA (2018) Draft genome of a high value tropical timber tree, Teak (Tectona grandis L. f): insights into SSR diversity, phylogeny and conservation. DNA Res 25(4):409–419

Zhang J, Guan W, Huang C, Hu Y, Chen Y, Guo J, Zhou C, Chen R, Du B, Zhu L, Huanhan D (2019a) Combining next-generation sequencing and single-molecule sequencing to explore brown plant hopper responses to contrasting genotypes of japonica rice. BMC Genomics 20(1):1–18

Zhang J, Yang E, He Q, Lin M, Zhou W, Pian R, Chen X (2019b). Genome-wide analysis of the WRKY gene family in drumstick (Moringa oleifera Lam.). Peer J 7: e7063

Zhang J, Yang E, He Q, Lin M, Zhou W, Pian R, Chen X (2019c) Genome-wide analysis of the WRKY gene family in drumstick (Moringa oleifera Lam.). Peer J 7: e7063

Zhao Q, Feng Q, Lu H, Li Y, Wang A, Tian Q, Zhan Q, Lu Y, Zhang L, Huang T, Wang Y (2018) Pan-genome analysis highlights the extent of genomic variation in cultivated and wild rice. Nat Genet 50 (2):278–284

Zhou Y, Zhao Z, Zhang Z, Fu M, Wu Y, Wang W (2019) Isoform sequencing provides insight into natural genetic diversity in maize. Plant Biotechnol J 17 (8):1473

Zulfiqar F, Casadesús A, Brockman H, Munné-Bosch S (2020) An overview of plant-based natural biostimulants for sustainable horticulture with a particular focus on moringa leaf extracts. Plant Sci 295:110194

Bioinformatics Tools and Moringa Genomics

12

Saranya Nallusamy and Jayakanthan Mannu

Abstract

Bioinformatics is an interdisciplinary field of biology, computer science, and mathematics. The advancement of high throughput genomics and proteomics technologies has produced large volume of genomics and proteomics data, which can be accessible from open databases. Exploring this big data could resolve many of the biological complexity. In this chapter, we discussed the role of bioinformatics in analyzing the genomics and proteomics data of Moringa.

12.1 Introduction

Moringa oleifera is a miracle tree in the ecosystem which is also called as drumstick tree/ben oil tree/horse radish tree or simply moringa. It serves as food (super food) with enriched nutritional value, livestock hood, agricultural applications, and possesses enormous medicinal properties (Matic et al. 2018). As an indigenous plant with Indian origin, it has been used as traditional medicine over centuries (Fahey 2005). All the plant parts like flower, leaf, root, seed, and stem are edible, and its phytochemicals and crude extract were proven to have antioxidant, anticancer, anti-inflammatory, antidiabetic, and antimicrobial activities against more than 300 human diseases (Anwar et al. 2007; Goyal et al. 2007).

With the advancement of high throughput omics technology, genomics and transcriptomics data of *Moringa oleifera* are made available in the public repositories. Bioinformatics studies of gene prediction, annotation, and pathway analysis provides insight on the biology of genes and their interaction mechanism that are required to understand the key biological and metabolic functions in moringa. Orthology analysis has shown its close evolutionary phylogenetic relation with *Carica papaya*, *Theobroma cocoa, Arabidopsis thaliana,* and *Vitis vinifera* (Tian et al. 2015; Pasha et al. 2020) compared to other species in viridiplantae kingdom.

Comparative genomics and transcriptomics studies provide massive knowledge on the similarities and differences between different plant species in the ecosystem. Moreover, biological databases and software are the key players that pave the scientists to explore new aspects in the way of metabolic and genetic engineering of moringa to produce value-added products in huge volumes. Apart from the huge medicinal value, it is also interesting to know the application of moringa gum in calico printing.

S. Nallusamy (✉) · J. Mannu
Department of Plant Molecular Biology and Bioinformatics, Centre for Plant Molecular Biology & Biotechnology, Tamil Nadu Agricultural University, Coimbatore, India
e-mail: saranya.n@tnau.ac.in

N. M. Boopathi et al. (eds.), *The Moringa Genome*, Compendium of Plant Genomes,
https://doi.org/10.1007/978-3-030-80956-0_12

Moringa oleifera being a reservoir of medicinal and nutritional properties was not completely characterized at the molecular and physiology level. There exists a lot of space to explore the mechanism of function and interaction that prevails within the *Moringa oleifera* genome.

12.2 Bioinformatics tools and databases

Bioinformatics is the interdisciplinary science in which informatics is applied in any biological data by means of computational tools and databases that renders scientific community the ease of data storage, access, interpretation, and analysis. High-throughput sequencing of any plant genome produces huge volume of data. Genomic databases like NCBI–Genome (Sayers et al. 2019), plant genomic database (http://www.plantgdb.org), Gramene (http://www.gramene.org/), Phytozome (Goodstein et al. 2012), and Ensemble plants (https://plants.ensembl.org/index.html) provides the genome sequence and associated information of plants.

These databases also contain tools integrated with their server, so that anyone can browse for the information regarding genes, chromosomes, markers, restriction sites, function, pathway, etc., without any restriction to access. Availability of genomic resources is the key to acquire molecular level genetic knowledge of an organism. Some of the mostly used tools and database information are provided in Table 12.1.

12.3 Pairwise Sequence Alignment

Most powerful tool that is being employed in bioinformatics is the Basic Local Alignment Search Tool (BLAST) (Johnson et al. 2008), which is a biological sequence comparison program based on the similarity. This tool compares the user-given biological sequences (nucleotide or protein sequences) to the sequences already stored in the databases and calculates the statistical significance of matches.

Match is found by pairwise aligning the user-given sequences to the database sequences. Information pertaining to function and evolutionary relationship could be retrieved by using this tool. BLOSUM (Blocks Substitution Matrix), PAM (Point Accepted Mutation) matrices are used for scoring the similarity between the sequences.

12.4 Multiple Sequence Alignment and Phylogenetic Tree Construction

To compare more than two sequences, unlike the pairwise similarity program, tools like CLUSTALW (Higgins and Sharp 1988), MAFFT (Katoh et al. 2019) are used. Multiple sequence alignment forms the basis for understanding the conserved and variable regions across the gene families. Other informations like motif, domain, signature, fingerprint, etc., are obtained by means of performing multiple sequence alignment.

Moreover, phylogenetic tree construction programs highly rely on the multiple sequence alignment programs to infer evolutionary relationship between the set of sequences. Programs like PHYLIP (Felsenstein 1993) and MEGA (Kumar et al. 2018) are mostly used for the phylogenetic tree construction.

Evolutionary information could be obtained in the form of gene tree or species tree based on the biological data available for study. Phylogenetic tree in simple can be compared with that of real tree which has branches, leaves, and root. Sequences with similar characteristics are grouped together in separate branches (internal node) and with the edge as the terminal node.

The phylogenetic tree can be rooted or unrooted depending on the ancestral sequence information availability. If ancestor or the primary sequences are present, there exists a root from which other sequences evolve by means of duplication or deletion or insertion or mutation. Methods such as clustering, Maximum likelihood, Maximum Parsimony, genetic algorithm, and Bayesian simulation are employed to construct the phylogenetic tree.

Table 12.1 List of prioritized bioinformatics databases and tools for *Moringa oleifera* genome and transcriptome analysis

S. no	Type	Databases/tools	Description	Weblink/References
1	Sequence databases and tools	**NCBI resources** Gene, Protein, Genome, GEO datasets, Bioproject, SRA experiment	NCBI provides public access to gene, protein, and genome and transcriptome data	https://www.ncbi.nlm.nih.gov/ (Sayers et al. 2019)
		Uniprot	Resource for the protein sequence and annotation data	https://www.uniprot.org/ (Bateman et al. 2020)
		Sequence Manipulation Suite	Tool serves the purpose of generating, formatting, and analyzing short DNA and protein sequences	https://www.bioinformatics.org/sms2/ (Stothard 2000)
		PFAM	Pfam is a database of protein families that includes their annotations and multiple sequence alignments generated using hidden Markov models	https://pfam.xfam.org/ (Finn et al. 2014)
		Motif Scan	Tool to find all the motif in a sequence	https://myhits.sib.swiss/cgi-bin/motif_scan (Hau et al. 2007)
		InterPro	InterPro performs functional analysis of proteins by classifying them into families and predicting domains and important sites. InterPro makes use of protein signature information from several different databases	http://www.ebi.ac.uk/interpro/ (Finn et al. 2017)
2	Genomics/transcriptomics tools	Cufflink	Transcriptome data analysis package Cufflinks assembles transcripts, estimates their abundances, and tests for differential expression and regulation in RNA-Seq samples	https://github.com/cole-trapnell-lab/cufflinks (Trapnell et al. 2012)
		SOAPdenovo2	Genome short read assembly program	https://github.com/aquaskyline/SOAPdenovo2 (Luo et al. 2015)
		OMICSBOX/Blast2GO	Genome Transcriptome and metagenomics data analysis package	Standalone commercial software (Götz et al. 2008; Workbench)
		CLC genomics workbench		
		HMMER	Hidden Markov Model can be employed to find the sequence similarity based on the profile of the aligned sequences. Build HMM profile can then be used to search large sequence databases to find related sequences, even those distantly related	http://hmmer.org/ (Eddy 2011)

(continued)

Table 12.1 (continued)

S. no	Type	Databases/tools	Description	Weblink/References
3	SSR Marker database and tools	Gramene SSR database	Plant SSR marker information	https://archive.gramene.org/markers/ (Tello-Ruiz et al. 2018)
		Krait	A robust and flexible tool for fast investigation of microsatellites in DNA sequences	https://github.com/lmdu/krait (Du et al. 2018)
		MISA	The MISA microsatellite finder (Thiel et al. 2003) is a tool for finding microsatellites in nucleotide sequences	MISA (Beier et al. 2017)
4	Phylogenetic analysis	PHYLIP	Phylogenetic tree construction software	https://evolution.genetics.washington.edu/phylip.html (Felsenstein 1993)
		MEGA (Molecular Evolutionary Genetics Analysis)	Enables the study of phylogenetic and evolutionary relationship in biological sequences. Includes statistical analysis packages	https://www.megasoftware.net/ (Kumar et al. 2018)
5	Sequence alignment	BLAST	Pairwise sequence alignment	https://blast.ncbi.nlm.nih.gov/ (Boratyn et al. 2013)
		CLUSTALW	Multiple sequence alignment	https://www.ebi.ac.uk/Tools/msa/clustalo/ (Higgins and Sharp 1988)
		MAFFT		https://mafft.cbrc.jp/alignment/server/ (Katoh et al. 2019)
6	Orthology analysis	Orthofinder	Software for analysis of phylogeny of orthogroups, gene duplication events in protein sequences	https://github.com/davidemms/OrthoFinder (Emms and Kelly 2019)
		Benchmarking Universal Single-Copy Orthologs (BUSCO)	Assessment of completeness of genome/transcritpome	http://busco.ezlab.org/ (Seppey et al. 2019)
		OrthoDB	Database of orthologs	(www.orthodb.org) (Kriventseva et al. 2019)
		OrthoMCL	Tool used for constructing orthologous groups across multiple eukaryotic taxa, using a Markov Cluster algorithm to group (putative) orthologs and paralogs	https://orthomcl.org/orthomcl/ (Li et al. 2003)

(continued)

Table 12.1 (continued)

S. no	Type	Databases/tools	Description	Weblink/References
7	Pathway database	KEGG pathway	Pathway maps representing the information of molecular interaction, reaction, and relation networks	https://www.genome.jp/kegg/pathway.html (Kanehisa et al. 2017)
		Reactome		https://reactome.org/ (Fabregat et al. 2018)
8	Protein–protein Interaction database and tools	StringDB	Database of known and predicted protein–protein interactions	https://string-db.org/ (Szklarczyk et al. 2020)
		Cytoscape	Software for visualizing molecular interaction networks and biological pathways and integrating these networks with annotations, gene expression profiles	https://cytoscape.org/
9	Non-coding RNA databases and tools	Coding Potential Calculator (CPC2)	Assessment of the coding and non-coding ability of RNA transcripts	http://cpc2.cbi.pku.edu.cn/ (Kang et al. 2017)
		Coding Potential Assessment Tool (CPAT)		http://lilab.research.bcm.edu/cpat/ (Wang et al. 2013)
		RFAM	The Rfam database stores the information of RNA sequence families such as the structural RNAs, non-coding RNA genes, and cis-regulatory elements	http://rfam.xfam.org/ (Kalvari et al. 2020)
		PmiREN (Plant miRNA ENcyclopedia)	Functional information of plant miRNA	http://www.pmiren.com/
		Noncode	Collection of Plant non-coding sequences	http://www.noncode.org/ (Xiyuan et al. 2017)
		CANTATAdb		http://cantata.amu.edu.pl/ (Szcześniak et al. 2016)
		psRNATarget	Prediction of target mRNA sequence for miRNA	http://plantgrn.noble.org/psRNATarget/ (Dai and Zhao 2011)

12.5 Genomics Tools

Genome/Transcriptome sequencing results in the generation of raw reads of large volume. These reads exist as an input data for the plethora of bioinformatics tools to carry out the steps such as alignment, annotation, and analysis. Quality checking and trimming of reads are the primary step before start of the alignment/assembly of raw reads.

FastQC (FastQC 2015; Andrews 2010) is the tool used for the quality checking of the raw reads based on the parameters such as GC content, duplication level, length distribution, etc. Trimming of raw reads is mainly performed to remove adapter sequences that may occur by chance in the sequencing procedure for which software like Trimmomatic (Bolger et al. 2014) is mostly used.

Next step is the assembly step which can either be reference-based and de novo-based. If reference genome is available, then the former methodology is employed, otherwise the latter is used. A number of assembly algorithms like Trinity Grabherr (Grabherr et al. 2011), STAR (Dobin et al. 2013), and SOAPdenovo (Luo et al. 2015) software are used for alignment/assembly process.

After assembly, gene prediction programs like AUGUSTUS (Stanke and Morgenstern 2005), FGENESH (Salamov and Solovyev 2000), GENSCAN (Burge and Karlin 1997) are used for predicting the coding part of the genome. Annotation is performed by similarity search against sequences of closely related species using BLAST, MAKER (Cantarel et al. 2008), BlastKOALA, and GhostKOALA tools (Kanehisa et al. 2016).

In this procedure, gene ontology terms such as Molecular Function, Cellular component, and Biological process are mapped based on their similarity with the already annotated sequences. This analysis provides the knowledge on its function, pathway, and their interaction mechanism. Other downstream analysis like orthology, non-coding region annotation is performed in order to understand gene duplication events, divergence that has occurred during the evolutionary period of time, and to decode the regulatory role of non-coding part of the genome.

System biology and synthetic biology studies have joined hand with genomic studies (Jamil et al. 2020) to engineer the biological system at the metabolic level, genetic level, or protein level mainly to increase the production of metabolites of pharmaceutical and medicinal value.

Comparative genomics studies involve the comparison of whole genomes of two or more species to understand the intron–exon organization, gene structure, and similarity/dissimilarity/conservation observed among the distribution of gene families. BLAST tool could be used for comparison of genomes. Genome level alignment and comparison also provides information on evolutionary events like speciation, duplication, and horizontal gene transfer events.

12.6 Moringa Genomics

First draft genome sequence of *Moringa oleifera Lam.* was first published in the year 2015 (Tian et al. 2015) from Yunnan Agricultural University, China. 19,465 annotated protein-coding genes were predicted from the 457× coverage DNA sequencing data for the *Moringa oleifera* sample. Based on the 17-mer frequency distribution, the estimated genome size was 315 Mb. Clustering analysis to understand the distribution of gene families among other plant species such as *Vitis vinifera*, *Cajanus cajan*, *Carica papaya,* and *Malus domestica* showed that these five different plant species possess similar numbers of gene families, with a core set of 10,215 shared genes. 198 single-copy gene families were found among the 12, 298 gene families reported in *Moringa oleifera*.

Second draft genome, as well as transcriptome data, was published by Chang et al. (2019) using *Moringa oleifera* sample grown at the World AgroForestry Center campus in Kenya. Genome size was observed to be 216.76 Mb. Transcription factors such as bHLH, NAC, ERF, MYB-related, C2H2, MYB, WRKY, bZIP, FAR1,

C3H, B3, G2-like, Trihelix, LBD, GRAS, M-type MADS, HDZIP, MIKC MADS, HSF, and GATA were found in abundance in comparison with other transcription factors.

Recently, RNA-sequencing enabled the analysis of gene expression samples from five different tissues (leaf, root, stem, seed, and flower) of *Moringa oleifera* plant (Bhagya variety) at the University of Agricultural Sciences, GKVK, Bangalore, India (Pasha et al. 2020). Pathway analysis was performed to understand the biosynthesis of secondary metabolites such as quercetin, kaempferol, benzylamine, and ursolic/oleanolic acid synthesized by *Moringa oleifera* genes which have profound medicinal values.

Being a drought-tolerant plant, stress-related transcription factors and enzymes related to production of metabolites of medicinal value were mostly expressed in the *Moringa oleifera* transcriptome leaf analysis. For example, pathway analysis showed the involvement of seven enzymes such as 4-Coumarate-CoA ligase (4CL), Chalcone synthase (CHS), Chalcone flavone isomerase (CHI), Flavonone 3-hydroxylase (F3H), Flavonol synthase (FLS), Tricin synthase (OMT), and Flavonoid 3′-monooxygenase (F3'H) in biosynthesis of anti-cancerous compound Quercetin.

Chloroplast genome of *Moringa oleifera* reported 131 genes (Lin et al. 2019) and was found to have a length of 160,600 bp with a large single-copy (LSC) region of 88,577 bp, a small single-copy (SSC) region of 18,883 bp, separated by two inverted repeat (IR) regions of 26,570 bp each. Phylogenetic analysis of 71 protein-coding sequences of 13 plant plastomes showed that *Moringa oleifera* is closest to *Carica papaya*.

WRKY transcription factors are well known for their role in plant development, signal transduction, and stress responses (Zhang et al. 2019). This gene family has been characterized in a genomic scale in *Moringa oleifera* through bioinformatics tools through the analysis of gene structures, motif analysis, conserved motifs, and phylogenetic tree construction. Fifty-four WRKY genes were identified through HMM profile search performed using WRKY domains.

Phylogenetic analysis showed its close relation with *Arabidopsis thaliana*. Also, analysis of commonly occurring cis-acting elements in WRKY promoter regions reported hormone responsive elements (ABRE, CGTCA motif, and TGACG motif), a drought stress responsive element (MBS), a heat stress responsive element (HSE), and four light responsive elements (Sp1, Box 4, G box, and GT1 motif), respectively.

Further expression profiling of WRKY genes reported its significance in various abiotic stress conditions such as under drought, salt, cold, and heat stresses. In a similar study of genome-wide analysis of trehalose-6-phosphate synthase (TPS) family, Group II *Moringa oleifera TPS* genes have evolved under relaxed purifying selection or positive selection. Further, group I *TPS* genes closely relate to reproductive development, and Group II *TPS* genes closely relate to high temperature resistance in leaves, stem, stem tip, and roots. Expression pattern of WRKY genes and TPS genes is experimentally validated using reverse transcription polymerase chain reaction (RT-PCR) and quantitative RT-PCR (qRT-PCR) experiments under different stress conditions.

12.7 Computational Identification of *Moringa oleifera* miRNA

Highly enriched nutritive and medicinal value of miracle tree is the repository of bioactive phytochemicals which has the potential to improve the health conditions of prevailing malnutrition observed among children from poorer section of the society and also pregnant women. Understanding the regulatory role of plant is very important to acquire the knowledge on the production of various phytochemicals under different stress conditions. MicroRNA (miRNA) is a small non-coding RNA of length about 22 nucleotides having an important role in regulating the gene expression at both post-transcriptional and translational level.

Moringa leaves and cold stressed callus (Pirrò et al. 2016) are characterized for the presence of conserved and novel microRNA families through RNA sequencing technology. Analysis using miRBase database and miRDeep2 tool predicted 431 conserved and 392 novel microRNAs, and it was confirmed using qRT-PCR analysis. Among the reported microRNAs, microRNA159 was majorly observed in leaf and callus, respectively. These miRNAs majorly targets the transcription factor that controls the plant growth, reproduction, and stress response.

Furthermore, plant microvesicles (MVs) possess similar features with that of mammalian exosomes, which are involved in cell–cell communication and miRNA transporters.

Moringa oleifera has shown enormous medicinal and pharmaceutical properties in a number of human diseases. In this context, ingestion of plant miRNA has shown regulation of gene activity in human which makes it a remarkable bioactive constituent for treating a number of human diseases. High-throughput sequencing of moringa seeds has reported miRNAs that has the potential to regulate the human gene expression at the post-transcriptional level (Pirrò et al. 2016). This regulation has the impact on exerting medicinal activity in a number of human diseases.

In silico analysis of miRNAs of moringa seeds for their contribution to medicinal value using MirCompare and combinatorial miRNA target prediction (COMIR) web tool identified *Moringa oleifera*–miR168a as a potential candidate for regulation of human genes. These genes are majorly involved in the cell–cycle regulation and p53 pathway. It is interesting that *SIRT1* (*Sirtuin*) gene was positively regulated by miR168a which was confirmed by transfection experiment.

12.8 Computational Screening of Potential Bioactive Compounds from *Moringa oleifera*

Computational screening of potential bioactive compounds against various biological protein targets for different diseases are gaining importance as complement to traditional drug design and discovery process nowadays. Success of this method depends on the identification of valid protein targets from the genomic region of the organism.

Complete genome sequence of the *Moringa oleifera* (Tian et al. 2018) gives the possibility of exploring various biological targets for drug design and discovery to treat various diseases. The minimum requirement of inputs for computational screening of chemical compounds is availability of three-dimensional structure of the target protein. This structure could be retrieved from protein structure database called Protein Data Bank (PDB) (www.rcsb.org), which contains experimentally determined three-dimensional structures of the proteins.

Comparative homology modeling approach could help us to model the three-dimensional structure of the protein in the case of non-availability of experimental structures. Similarly, structure of phytochemicals can be retrieved from PubChem (https://pubchem.ncbi.nlm.nih.gov) database, which is an open database of chemical compounds maintained by National Institute of Health.

In recent years, many of the phytochemicals of *Moringa oleifera* have been virtually screened against various disease targets such as Diabetes Mellitus, different cancers, hypertension, COVID-19, antimicrobial activity, antioxidant defense systems, and HIV.

12.8.1 Diabetes Mellitus

Phytochemicals of *Moringa oleifera* such as anthraquinone, 2-phenylchromenylium (Anthocyanins), hemlock tannin, sitogluside (glycoside), and A-phenolic steroid were reported as potential therapeutic agents against mutated insulin receptor using molecular docking approach (Zainab et al. 2020).

The strategy of toxicity screening, checking for Lipinski rule violation, and pharmacophore generation has been carried out along with docking study for the phytochemicals of *Moringa oleifera* in their study. Similarly, Yang et al. (2014) have performed virtual screening, docking, ADME prediction, and in-vitro analysis for the phytochemicals of *Moringa oleifera* to identify potential compounds for diabetics.

A total of 111 phytochemicals were screened in their analysis against Potential Dipeptidyl Peptidase (DPP)-IV, and it was reported O-Ethyl-4-[(α-L-rhamnosyloxy)-benzyl] carbamate has the activity with half-maximal inhibitory concentration [IC50] = 798 nM.

12.8.2 Carcinoma

Crude ethanolic extract (HF-CEE) of *Moringa oleifera* seeds reported for their inhibitory action on MCF7 breast cancer cell growth (Mansour et al. 2019). Methylelaidate, a phytochemical of *Moringa oleifera,* was tested for their binding efficiency with Bax and MDM2 apoptotic proteins using AutoDock. This showed a stable interaction with the target proteins and also has the potential of drug-like properties according to the Lipinski's rule of 5 Adebayo (Adebayo et al. 2018).

Another phytochemical, quinicacid, was reported for their better pharmacokinetic properties and suitable for further drug discovery and development cycle to control prostate cancer cell growth (Inbathamizh and Padmini 2013).

12.8.3 Hypertension

Some of the phytochemicals of *Moringa oleifera* such as Niazicin-A, Niazimin-A, and Niaziminin-B were tested for their binding efficiency with Angiotensin-converting enzyme (ACE) using AutoDockVina and for pharmacokinetics activity using ADME-Toxicity prediction. The above compounds showed good binding energy compared to the reference drug molecules such as Captopril and Enalapril (Khan et al. 2019; Aktar et al. 2019). They have reported that leaves methanolic extract (MOLME) of *Moringa oleifera* showed inhibitory activity against Angiotensin-converting enzyme using spectrophotometric method. It is also reported that substrate hippuryl-L-histidyl-L-leucine (HHL) inhibits ACE with an IC50 value of 226.37 μg/ml, in comparison to reference compound, captropril, which shows IC50 value of 0.0289 μM.

12.8.4 COVID-19

Currently, there is no potential drug or vaccine developed to target SARS-CoV-2 virus. It is mandatory to identify a drug to handle the pandemic situation prevailing worldwide. The multiple protein targets of SARS-CoV-2 could be accessible from publicly available genomic and proteomic databases. Phytochemical compounds of *Moringa oleifera* were virtually screened using AutoDockVinato to discover novel lead compounds against main protease (Mpro) and RNA-dependent RNA polymerase (RdRp) to treat COVID-19.

The compounds were also screened for drug-likeness properties using Swiss ADME. The scientists have reported that the compounds kaempferol, pterygospermin, morphine, and quercetin made a stable interaction with Mpro and RdRp target proteins. These compounds could be taken as potential lead molecules for further evaluation to treat COVID-19 (Shaji 2020).

12.9 Future Prospects and Applications

In the present scenario of increase in the pandemic and life threating diseases, every human is in need of super food like *Moringa oleifera* which eradicates malnutrition as well as increase immunity and longevity of life. It is very important to understand the molecular mechanism of biomolecules and their pathways to have future success stories on metabolomic and genetic engineering which will help in the production of value-added products in the pharmaceutical research. Enormous research studies are required to completely link the genes, proteins, metabolites, non-coding genes, and their interaction mechanisms in a concurrent way.

References

Adebayo IA, Arsad H, Samian MRJPM (2018) Methyl elaidate: A major compound of potential anticancer extract of Moringa oleifera seeds binds with bax and MDM2 (p53 inhibitor) In silico. Pharmacogn Mag 14 (59):554

Aktar S, Das PK, Asha SY, Siddika MA, Islam F, Khanam JA, Rakib MA (2019) Moringa oleifera leaves methanolic extract inhibits angiotensin converting enzyme activity in vitro which ameliorates hypertension. J Adv Biotechnol Exp Ther 2(2):73–77

Andrews S (2010) FastQC: a quality control tool for high throughput sequence data. https://www.bioinformatics.babraham.ac.uk/projects/fastqc/

Anwar F, Latif S, Ashraf M, Gilani AH (2007) Moringa oleifera: a food plant with multiple medicinal uses. Phytother Res Int J Devoted Pharmacol Toxicol Eval Nat Prod Deriv 21(1):17–25

Bateman A, Martin M-J, Orchard S, Magrane M, Agivetova R, Ahmad S, Alpi E, Bowler-Barnett EH, Britto R, Bursteinas B (2020) UniProt: the universal protein knowledgebase in 2021. Nucl Acids Res 49 (D1):D480–D489

Beier S, Thiel T, Münch T, Scholz U, Mascher M (2017) MISA-web: a web server for microsatellite prediction. Bioinformatics 33(16):2583–2585

Bolger AM, Lohse M, Usadel B (2014) Trimmomatic: a flexible trimmer for Illumina sequence data. Bioinformatics 30(15):2114–2120

Boratyn GM, Camacho C, Cooper PS, Coulouris G, Fong A, Ma N, Madden TL, Matten WT, McGinnis SD, Merezhuk Y (2013) BLAST: a more efficient report with usability improvements. Nucl Acids Res 41(W1):W29–W33

Burge C, Karlin S (1997) Prediction of complete gene structures in human genomic DNA. J Mol Biol 268 (1):78–94

Cantarel BL, Korf I, Robb SM, Parra G, Ross E, Moore B, Holt C, Alvarado AS, Yandell M (2008) MAKER: an easy-to-use annotation pipeline designed for emerging model organism genomes. Genome Res 18(1):188–196

Chang Y, Liu H, Liu M, Liao X, Sahu SK, Fu Y, Song B, Cheng S, Kariba R, Muthemba S and Hendre PS (2019) The draft genomes of five agriculturally important African orphan crops GigaScience 8 (3):p. giy152.

Dai X, Zhao PX (2011) psRNATarget: a plant small RNA target analysis server. Nucl Acids Res 39(suppl_2): W155–W159

Dobin A, Davis CA, Schlesinger F, Drenkow J, Zaleski C, Jha S, Batut P, Chaisson M, Gingeras TR (2013) STAR: ultrafast universal RNA-seq aligner. Bioinformatics 29(1):15–21

Du L, Zhang C, Liu Q, Zhang X, Yue B (2018) Krait: an ultrafast tool for genome-wide survey of microsatellites and primer design. Bioinformatics 34(4):681–683

Eddy SR. Accelerated Profile HMM Searches. PLoS Comput Biol. 2011 Oct;7(10):e1002195. https://doi.org/10.1371/journal.pcbi.1002195

Emms DM, Kelly S (2019) OrthoFinder: phylogenetic orthology inference for comparative genomics. Genome Biol 20(1):1–14

Fabregat A, Jupe S, Matthews L, Sidiropoulos K, Gillespie M, Garapati P, Haw R, Jassal B, Korninger F, May B (2018) The reactome pathway knowledgebase. Nucl Acids Res 46(D1):D649–D655

Fahey JW (2005) Moringa oleifera: a review of the medical evidence for its nutritional, therapeutic, and prophylactic properties. Part 1. Trees Life J 1(5):1–15

FastQC (2015). https://qubeshub.org/resources/fastqc

Felsenstein J (1993) PHYLIP (phylogeny inference package), version 3.5 c. Joseph Felsenstein

Finn RD, Attwood TK, Babbitt PC, Bateman A, Bork P, Bridge AJ, Chang H-Y, Dosztányi Z, El-Gebali S, Fraser M (2017) InterPro in 2017—beyond protein family and domain annotations Nucl Acids Res 45 (D1):D190–D199

Finn RD, Bateman A, Clements J, Coggill P, Eberhardt RY, Eddy SR, Heger A, Hetherington K, Holm L, Mistry J (2014) Pfam: the protein families database. Nucl Acids Res 42(D1):D222–D230

Goodstein DM, Shu S, Howson R, Neupane R, Hayes RD, Fazo J, Mitros T, Dirks W, Hellsten U, Putnam N (2012) Phytozome: a comparative platform for green plant genomics. Nucl Acids Res 40(D1): D1178–D1186

Götz S, García-Gómez JM, Terol J, Williams TD, Nagaraj SH, Nueda MJ, Robles M, Talón M, Dopazo J, Conesa A (2008) High-throughput functional annotation and data mining with the Blast2GO suite. Nucl Acids Res 36(10):3420–3435

Goyal BR, Agrawal BB, Goyal RK, Mehta AA (2007) Phyto-pharmacology of Moringa oleifera Lam.—an overview. Nat Prod Radiance 6 (4)

Grabherr MG, Haas BJ, Yassour M, Levin JZ, Thompson DA, Amit I, Adiconis X, Fan L, Raychowdhury R, Zeng Q (2011) Full-length transcriptome assembly from RNA-Seq data without a reference genome. Nat Biotechnol 29(7):644–652

Hau J, Muller M, Pagni M (2007) HitKeeper, a generic software package for hit list management. Source Code Biol Med 2(1):1–8

Higgins DG, Sharp PM (1988) CLUSTAL: a package for performing multiple sequence alignment on a microcomputer. Gene 73(1):237–244

https://doi.org/10.1007/s00122-002-1031-0

Inbathamizh L, Padmini E (2013) Quinic acid as a potent drug candidate for prostate cancer–a comparative pharmacokinetic approach. Asian J Pharm Clin Res 6(4):106–112

Jamil IN, Remali J, Azizan KA, Muhammad NAN, Arita M, Goh H-H, Aizat WM (2020) Systematic multi-omics integration (MOI) approach in plant systems biology. Front Plant Sci 11

Johnson M, Zaretskaya I, Raytselis Y, Merezhuk Y, McGinnis S, Madden TL (2008) NCBI BLAST: a better web interface. Nucl Acids Res 36(suppl_2): W5–W9

Kalvari I, Nawrocki EP, Ontiveros-Palacios N, Argasinska J, Lamkiewicz K, Marz M, Griffiths-Jones S, Toffano-Nioche C, Gautheret D, Weinberg Z, Rivas E (2021) Rfam 14: expanded coverage of metagenomic, viral and microRNA families Nucleic Acids Research49(D1):D192-200

Kanehisa M, Furumichi M, Tanabe M, Sato Y, Morishima K (2017) KEGG: new perspectives on genomes, pathways, diseases and drugs. Nucl Acids Res 45(D1):D353–D361

Kanehisa M, Sato Y, Morishima K (2016) BlastKOALA and GhostKOALA: KEGG tools for functional characterization of genome and metagenome sequences. J Mol Biol 428(4):726–731

Kang Y-J, Yang D-C, Kong L, Hou M, Meng Y-Q, Wei L, Gao G (2017) CPC2: a fast and accurate coding potential calculator based on sequence intrinsic features. Nucl Acids Res 45(W1):W12–W16

Katoh K, Rozewicki J, Yamada KD (2019) MAFFT online service: multiple sequence alignment, interactive sequence choice and visualization. Brief Bioinform 20(4):1160–1166. https://doi.org/10.1093/bib/bbx108

Khan H, Jaiswal V, Kulshreshtha S, Khan A (2019) Potential angiotensin converting enzyme inhibitors from Moringa oleifera. Recent Pat Biotechnol 13 (3):239–248

Kriventseva EV, Kuznetsov D, Tegenfeldt F, Manni M, Dias R, Simão FA, Zdobnov EM (2019) OrthoDB v10: sampling the diversity of animal, plant, fungal, protist, bacterial and viral genomes for evolutionary and functional annotations of orthologs. Nucl Acids Res 47(D1):D807–D811

Kumar S, Stecher G, Li M, Knyaz C, Tamura K (2018) MEGA X: molecular evolutionary genetics analysis across computing platforms. Mol Biol Evol 35 (6):1547–1549

Li L, Stoeckert CJ, Roos DS (2003) OrthoMCL: identification of ortholog groups for eukaryotic genomes. Genome Res 13(9):2178–2189

Lin W, Dai S, Chen Y, Zhou Y, Liu X (2019) The complete chloroplast genome sequence of Moringa oleifera Lam.(Moringaceae). Mitochondrial DNA Part B 4 (2):4094–4095

Luo R, Liu B, Xie Y, Li Z, Huang W, Yuan J, He G, Chen Y, Pan Q, Liu Y (2015) SOAPdenovo2: an empirically improved memory-efficient short-read de novo assembler. GigaScience 4(1):s13742-13015-10069-13742

Mansour M, Mohamed MF, Elhalwagi A, El-Itriby HA, Shawki HH, Abdelhamid IA (2019) Moringa peregrina leaves extracts induce apoptosis and cell cycle arrest of hepatocellular carcinoma. BioMed research international 1;2019.

Matic I, Guidi A, Kenzo M, Mattei M, Galgani A (2018) Investigation of medicinal plants traditionally used as dietary supplements: A review on *Moringa oleifera* J Public Health Afr. 9(3):841.

Pasha SN, Shafi KM, Joshi AG, Meenakshi I, Harini K, Mahita J, Sajeevan RS, Karpe SD, Ghosh P, Nitish S (2020) The transcriptome enables the identification of candidate genes behind medicinal value of Drumstick tree (Moringa oleifera). Genomics 112(1):621–628

Pirrò S, Zanella L, Kenzo M, Montesano C, Minutolo A, Potestà M, Sobze MS, Canini A, Cirilli M, Muleo R (2016) MicroRNA from Moringa oleifera: identification by high throughput sequencing and their potential contribution to plant medicinal value. PLoS One 11 (3):e0149495

QIAGEN CLC Genomics Workbench 20.0 (https://digitalinsights.qiagen.com/)

Salamov AA, Solovyev VV (2000) Ab initio gene finding in Drosophila genomic DNA. Genome Res 10(4):516–522

Sayers EW, Agarwala R, Bolton EE, Brister JR, Canese K, Clark K, Connor R, Fiorini N, Funk K, Hefferon T, Holmes JB (2019) Database resources of the national center for biotechnology information Nucleic acids research.47(Database issue):D23.

Seppey M, Manni M, Zdobnov EM (2019) BUSCO: assessing genome assembly and annotation completeness. In: Gene prediction. Springer, pp 227–245

Shaji D (2020) Computational Identification of Drug Lead Compounds for COVID-19 from Moringa Oleifera. ChemRxiv. Cambridge: Cambridge Open Engage; This content is a preprint and has not been peer-reviewed.

Stanke M, Morgenstern B (2005) AUGUSTUS: a web server for gene prediction in eukaryotes that allows user-defined constraints. Nucl Acids Res 33(suppl_2): W465-7.

Stothard P (2000) The sequence manipulation suite: JavaScript programs for analyzing and formatting protein and DNA sequences. Biotechniques 28 (6):1102–1104

Szcześniak MW, Rosikiewicz W, Makałowska I (2016) CANTATAdb: a collection of plant long non-coding RNAs. Plant Cell Physiol 57(1):e8–e8

Szklarczyk D, Gable AL, Nastou KC, Lyon D, Kirsch R, Pyysalo S, Doncheva NT, Legeay M, Fang T, Bork P

(2020) The STRING database in 2021: customizable protein–protein networks, and functional characterization of user-uploaded gene/measurement sets. Nucl Acids Res

Tello-Ruiz MK, Naithani S, Stein JC, Gupta P, Campbell M, Olson A, Wei S, Preece J, Geniza MJ, Jiao Y (2018) Gramene 2018: unifying comparative genomics and pathway resources for plant research. Nucl Acids Res 46(D1):D1181–D1189

Tian W, Chen C, Lei X, Zhao J, Liang J (2018) CASTp 3.0: computed atlas of surface topography of proteins. Nucl Acids Res 46(W1):W363–W367. https://doi.org/10.1093/nar/gky473

Tian Y, Zeng Y, Zhang J, Yang C, Yan L, Wang X, Shi C, Xie J, Dai T, Peng L, Huan YZ (2015) High quality reference genome of drumstick tree (Moringa oleifera Lam.), a potential perennial crop Science China Life Sciences 58(7):627-38.

Trapnell C, Roberts A, Goff L, Pertea G, Kim D, Kelley DR, Pimentel H, Salzberg SL, Rinn JL, Pachter L (2012) Differential gene and transcript expression analysis of RNA-seq experiments with TopHat and Cufflinks. Nat Protoc 7(3):562–578

Wang L, Park HJ, Dasari S, Wang S, Kocher J-P, Li W (2013) CPAT: Coding-Potential Assessment Tool using an alignment-free logistic regression model. Nucl Acids Res 41(6):e74–e74

Xiyuan L, Dechao B, Liang S, Yang W, Shuangsang F, Hui L, Haitao L, Chunlong L, Wenzheng F, Runsheng C and Yi Z (2017) Using the NONCODE database resource Current protocols in bioinformatics 58(1):12-16.

Yang X, Chen X, Bian G, Tu J, Xing Y, Wang Y, Chen Z (2014) Proteolytic processing, deubiquitinase and interferon antagonist activities of Middle East respiratory syndrome coronavirus papain-like protease. J Gen Virol 95(3):614–626. https://doi.org/10.1099/vir.0.059014-0

Zainab B, Ayaz Z, Alwahibi MS, Khan S, Rizwana H, Soliman DW, Alawaad A, Abbasi AM (2020) In-silico elucidation of Moringa oleifera phytochemicals against diabetes mellitus. Saudi J Biol Sci 27(9):2299–2307

Zhang J, Yang E, He Q, Lin M, Zhou W, Pian R, Chen X (2019) Genome-wide analysis of the WRKY gene family in drumstick (*Moringa oleifera* Lam.). PeerJ 7: e7063

Biogenic Nanomaterials Using Moringa and Their Applications

13

M. Kannan, K. Govindaraju, K. Elango, M. Kalyanasundaram, and A. Lakshmanan

Abstract

Nanotechnology offers many potential advantages for agriculture such as enhancement of food quality and safety, reduction of agricultural inputs, enrichment of absorbing nanoscale nutrients from the soil, etc. Such nanotechnological applications reduce the amount of spread of chemicals, minimize nutrient losses in fertilization and increase the yield through pest and nutrient management. Besides, it also provides novel nanotools for rapid disease diagnostic and controlling measures using nanopesticides and nano-sensors for monitoring soil quality of agricultural field. This chapter covers the current status and challenges in the area of nanotechnology in the improvement of agriculture with a special emphasis given to Moringa.

M. Kannan (✉) · A. Lakshmanan
Department of Nano Science and Technology, Tamil Nadu Agricultural University, Coimbatore 641003, India
e-mail: nanotech@tnau.ac.in

K. Govindaraju
Centre for Ocean Research (DST-FIST Sponsored Centre), Sathyabama Institute of Science and Technology, Chennai 600119, India

K. Elango
Horticultural Research Station, Tamil Nadu Agricultural University, Kodaikanal 624103, India

K. Elango
Department of Agricultural Entomology, Tamil Nadu Agricultural University, Coimbatore 641003, India

M. Kalyanasundaram
Department of Nano Science and Technology, Tamil Nadu Agricultural University, Coimbatore 641003, India

13.1 Introduction

In recent days, great technological developments in the area of research and technology have been initiated by nano science and technology. Nanotechnology is the research and application of small structures that can be seen in all areas such as chemistry, biology, physics, material sciences and engineering. The idea of nanotechnology originated in the ninth century. For the first time in 1959, Richard Feynman gave a talk on the concept of nanotechnology and described molecular machines built with atomic precision, where he discussed nanoparticles and said, 'There's plenty of space at the bottom'. Nanotechnology is an increasingly growing research area that is an interdisciplinary field of science and technology that expands the scope of cell-level investment and control between synthetic material and the biological system (Sinha et al. 2009). Inorganic nanomaterials play an important role in various applications in different fields, such as optics, electronics, catalysis, drug delivery, atmosphere, environmental and biomedicine, in particular biomedicine, because of their special chemical,

N. M. Boopathi et al. (eds.), *The Moringa Genome*, Compendium of Plant Genomes,
https://doi.org/10.1007/978-3-030-80956-0_13

electrical magnetic and physical properties (Guo and Wang 2011). Different metal and metal oxide nanoparticles synthesized by biological methods have been shown to be useful in different applications over the last two decades. In addition, these biological approaches satisfy the green chemistry approach requirements defined as the design of chemical products and processes that avoid the use and processing of toxic/hazardous chemicals (Govindaraju and Tamilselvan 2020). In the last few decades, due to their cost-effective, non-toxic and eco-friendly nature, biological synthesis of nanomaterials using different bio-resources has gained a lot of interest in the field of nanotechnology. Researchers have turned their attention to the synthesis of nanoparticles, motivated by the abilities of green methodologies, using diverse biological materials such as actinomycetes (Manimaran and Kannabiran 2017), bacteria (Iravani 2014; Mukherjee and Nethi 2019; Ali et al. 2019), Cynobacteria (Hamouda et al. 2019), fungi (Dorcheh and Vahabi 2016; Guilger-Casagrande and de Lima 2019), plants and plant compounds (Govindaraju et al. 2011; Ashokkumar et al. 2014; Khaleel Basha et al. 2010; Venkatachalam et al. 2013; Uma Suganya et al. 2016a, b; Prabhu et al., 2014; Govindaraju et al. 2020a), seaweeds (Singaravelu et al., 2007; Govindaraju et al., 2009; Govindaraju et al. 2020b, c; Itroutwar et al. 2020), yeasts (Moghaddam et al., 2015; Eugenio et al., 2016) and even viruses (Thangavelu et al., 2020).

Among the various bio-resources, plants are also being used as a 'green bio-factory' for the production of metal and metal oxide nanomaterials. Further, plant-based genre of bio-reductants, have more advantages due to their high metal ion reduction, low cost and easy availability.

Moringa oleifera, originally discovered in India, is a member of the Moringaceae family and is generally referred as Drumstick tree, the Miracle tree or the Horseradish tree. There are 13 species, but *Moringa oleifera* is the most grown with a height of between 5 and 10 m. As the tree grows quickly and has drought tolerance properties, it can be cultivated in tropical, subtropical and arid regions of the world. It is exceptional and has tremendous promise, highlighted by the National Institute of Health, Bethesda, Maryland. The plant is highly regarded since nearly all pieces are used as a food source, as well as in the conventional treatment of various diseases and to encourage good health. These sections include but are not limited to the leaves, flowers, seed pods, nuts, stems, bark and gum. It is used for the prevention of bronchitis infections and fever. Further, it can also be used as antioxidant, antimicrobial, anti-diabetic, neuroprotective, cardioprotective and anti-inflammatory effects (Madi et al. 2016; Tiloke et al. 2018). In this sense, the approach to green chemistry for the synthesis of nanomaterials of metal and metal oxides by using various sections of *Moringa* is now under tremendous investigation and has thus become a groundbreaking gleam in the research arena.

13.2 Synthesis of Silver Nanoparticles and Its Applications

Due to its distinct properties such as catalytic activity, chemically stable, good conductivity and potent antimicrobial activity, silver nanoparticles have attracted and demanding research interest in the field of nanotechnology (Setua et al. 2007; Govindaraju et al. 2009). It is used as a substratum for surface-enhanced spectroscopy because of its colloidal nature, as it partly requires electrical conducting surface. Silver was used as an antimicrobial agent in this era. In order to increase the treatment of antibiotic resistance caused by the misuse of antibiotics, the recent focus is on silver nanoparticle synthesis (Panaek et al. 2006; Sandbhy et al. 2006). Several hypotheses have been identified for the antimicrobial activity of silver nanoparticles. The capacity of the silver nanoparticle inactivates the bacterial enzyme by releasing ionic silver that inactivates the thiol groups. Bacterial DNA replication, damage to cell cytoplasm, depletion of ATP levels and ultimately cell death are mediated by these silver ions (Feng et al. 2000; Uma Suganya et al. 2015a, b). As a nanoparticle of distinct surface-to-volume ratio, silver nanoparticle increases surface-to-surface interaction with bacterial cells that facilitate the dissolution of silver ion and enhance the efficacy of bacterial ion (Stobie et al. 2008). Among the various methods (physical, chemical) for synthesis of silver nanoparticles, biological methods particularly plant-based synthesis have highly stable and potent antimicrobial activity. Silver nanoparticles biosynthesis using *Moringa* leaf extracts has shown that this process generates a very sharp distribution. Further, it has shown that nonlinear optical absorption and optical limiting phenomena with an output comparable to particles prepared by other pathways can be added to the particles synthesized through this process (Sathyavathi et al. 2010). Prasad and Elumalai (2011) reported that the aqueous leaf extract of *Moringa* has been used for synthesizing spherical-shaped (average size 57 nm) silver nanoparticles and their antibacterial activity against *S. aureus, E. coli, K. pneumonia* and *B. cereus* and anti-fungal activity against *C. albicans, C. tropicalis* and *C. krusei.* Silver nanoparticles mediated by *Moringa* stem bark extract show excellent anti-cancer activity against the HeLa cell type. Green synthesized silver nanoparticles were stable and their modes of action were found to increase the generation of ROS and its subsequent action to inhibit cell replication by induction of apoptosis (Vasanth et al. 2014). Sunlight irradiation induced green synthesized silver nanoparticles using leaf extracts of Moringa and its antimicrobial potential against gram positive and gram negative bacterial species (*Staphylococcus aureus, Enterococcus faecalis, Escherichia coli, Pseudomonas aeruginosa* and *Klebsiella pneumonia* (Moodley et al. 2018)). Kalugendo and Kousalya (2019) reported the synthesis of silver nanoparticles using seed extracts of *Moringa* and their antibacterial activity against methicillin-resistant *Staphylococcus aureus*. Shousha et al. (2019) reported that Moringa leaves phyto-chemicals mediated synthesized silver nanoparticles showed potent in vitro antioxidant and cytotoxic activity. Recently, biogenic preparation and physico-chemical characterization of silver nanoparticles using *Moringa* flower extract and assessment of antimicrobial and heavy metal sensing properties have been studied (Bindhua et al. 2020).

13.3 Synthesis of ZnO and ZnS Nanoparticles and Its Applications

Semiconductor nanomaterials, particularly ZnO/ZnS wide band gap optically transparent semiconductor, can be widely used in photonics, optics and optoelectronics with diverse applications ranging from optical coatings, transistors, sensors, optoelectronic devices, drug delivery and agricultural applications. Bio-production and characterization of ZnO nanoparticles have been carried out using leaf, stem segments, flowers and fruit pods of the medicinal plant *Moringa* (Manokari and Shekhawat 2016). Preparation of ZnO nanoparticles was done using *Moringa* leaves as natural precursor via co-precipitation technique, and the nanoparticles were characterized using various spectroscopy and microscopy tools. Further, ZnO has been used for photocatalytic degradation of titan yellow dye and antimicrobial activity against *Bacillus subtilis* and *Escherichia coli.* Biosynthesized ZnS nanoparticles using the moringa leaves extract and its optical, dielectric, electronic property were studied. Morphology of ZnS was found to be mostly spherical in shape with average diameter $\sim$30 nm (Sur and Ankamwar 2016).

13.4 Synthesis of Iron Nanoparticles and Its Applications

Katata-Seru et al. (2018) reported the biogenic preparation of iron nanoparticles using leaf and seed extracts of *Moringa* and their application in removal of nitrate from waste water and antibacterial activity against *Escherichia coli.* The removal percentage of nitrate is increased with a decrease in pH. Further, iron nanoparticles have dual properties of coagulant and antibacterial activities, which is ideal for treating contaminated water. Magnetic iron oxide nanoparticles were prepared and surface modified with *Moringa* seed proteins and characterization of functional groups responsible for adsorption, stability, morphology and surface interaction of nanoparticles was done. Batch adsorption technique has been used for recovery of precious metal ions Au(III), Pd(II) and Pt(IV) from aqueous solution (Amuanyena et al. 2019). Recently, Oliveira et al. (2020) reported that magnetite nanoparticles were prepared using co-precipitation technique and surface functionalized with *Moringa* seeds extract for magnetic coagulant MO-Fe_3O_4 application in effluent treatment plant.

13.5 Synthesis of Vanadium Nanoparticles and Its Applications

Recently, nanostructured vanadium compounds have attracted much interest due to their chemical and physical properties and their potential applications in catalysis, sensors, electrochemical capacitors, solar cells, biomedical devices (REF). Vanadium nanoparticles were synthesized by biogenic approach wherein *Moringa* leaf extract used as reducing and stabilizing agent. Antimicrobial activity of the biogenic vanadium nanoparticle was tested against bacteria and fungi.

13.6 Synthesis of Gold Nanoparticles

Gold nanoparticles have been used in a wide range of applications over the past few decades, such as biosensors, catalysis, drug delivery, biomedical and environmental applications. Gold core is inherently inert, biocompatible and very less toxic in nature due to its chemical stability and special properties, and hence, making it a perfect starting point for carrier construction. Nanoscale gold can be conveniently processed with controlled dispersity with a broad variety of core sizes, i.e. 1–150 nm. Different chemical and physical methods have been produced for nanoscale gold synthesis, but due to the use of a number of poisonous chemicals and high temperature in the preparation, these approaches have been found to be hazardous and detrimental to biological and environmental applications. Due to their easily reducible, quick, cost-effective, less toxic and environmentally safe nature, biological preparation of nanoscale gold is becoming dominant. Anirban et al. (2013) reported the *Moringa* leaf extract mediated green synthesis of spherical-shaped gold nanoparticles of size ranging between 20 and 60 nm at room temperature. Similarly, Ponnanikajamideen et al. (2016) used moringa gum for biological synthesis of gold nanoparticles for effective antibacterial activity.

13.7 Synthesis of Bismuth Nanoparticles and Its Applications

Das et al. (2020) reported the green synthesis of phyto-chemical encapsulated bismuth nanoparticles using a hydro-alcoholic extract of *M. oleifera* leaves. The size of the synthesized bismuth nanoparticles is in the range of 40–57 nm with amorphous morphology. Further, DPPH and phosphomolybdate assays revealed that the *M. oleifera* leaves extract and the synthesized bismuth nanoparticles possess antioxidant properties. Also, it was demonstrated that the *M. oleifera* leaves extract and the synthesized bismuth nanoparticles exert potent antibacterial activity against *Escherichia coli, Klebsiella pneumoniae, Staphylococcus aureus* and *Enterococcus faecalis* and anti-fungal activity against *Aspergillus niger, Aspergillus flavus, Candida albicans* and *Candida glabrata.*

13.8 Synthesis of NiO Nanoparticles and Its Applications

With several characteristics, such as surface area to volume ratio, electro-optical, magneto-optical, chemical and mechanical properties, nanoparticles differ from the bulk, serving as an effective instrument in this sense in the fight against bacteria. Due to high chemical stability, electro catalysis, super conductance characteristics and electron transfer capacity, NiO nanoparticles have attracted a lot of attention in recent study. Nickel oxide has a band gap of 3.6–4.0 eV and is a p-type semiconductor metal oxide. In addition, this paper documents that the NiO nanoparticles prepared from the green approach demonstrate improved cytotoxicity and antibacterial action.

Figure 13.1 describes the procedure involved in the development of nanoparticles using different parts of Moringa plant. Table 13.1 illustrates different applications of nanomaterials including development of nonlinear optics, nanomaterials with antimicrobial, antiviral and antioxidant capacity that were prepared using Moringa plant parts.

13.9 Future Prospects

Agriculture rapidly adapts newer technologies, modernization and increased use of nanochemicals. This warrant government policies need to be tailored to acclimatize these adaptations in order to maximize the production in agriculture and food industry, as nanotechnology possesses incredibly unique property and role in food supply chain (including crop production, use of agro-chemicals such as nanofertilizers, nanopesticides and nanoherbicides, precision farming techniques, intelligent feed, enhancement of food texture and quality, bioavailability/nutrient values, packaging and labeling). From the above, it has been shown that Moringa has played several roles in formulating different nanotechnological strategies, and these areas may need more attention in near future to take the Moringa-based nanoproducts to the next level.

Fig. 13.1 Formation of nanoparticles using different Moringa plant parts

Table 13.1 Applications of nanotechnology that have employed Moringa plant parts

Moringa Plant parts used for synthesis	Nanomaterials	Size	Shape	Application	References
Leaves	Silver nanoparticles	46 nm	Spherical in shape	Nonlinear optics	Sathyavathi et al. (2010)
		57 nm	Spherical morphology	Antimicrobial activity	Prasad and Elumalai (2011)
		30 nm	Spherical morphology	Antimicrobial activity	Nilanjana et al. (2014)
		1–56.9 nm; 2–448.1 nm; 3–4705.0 nm	Polydisperse	Antimicrobial efficacy in packing materials	Narwade et al. (2018)
		5 and 10 nm	Irregular in shape	In vitro cytotoxicity and antioxidant activity	Shousha et al. (2019)
		9 and 11 nm	Spherical in shape	Antimicrobial activity	Moodley et al. (2018)
	Iron nanoparticles	3.4 and 7.4 nm	Spherical in shape	Removal of nitrate from water and antibacterial activity	Katata-Seru et al. (2018)
	ZnO nanoparticles	–	–	–	Manokari and Shekhawat (2016)
		52 nm	Hexagonal wurtzite structure	Photocatalytic and Antibacterial Activity	Pal et al. (2018)
	Vanadium nanoparticles	100 nm	Spherical in shape	Antimicrobial Activity	Aliyu et al. (2017)
	Copper nanoparticles	35.8–49.2 nm	Spherical in shape	Antioxidant and Antimicrobial Activities	Das et al. (2020)
	Bismuth nanoparticles	40.4–57.8 nm	Amorphous morphology	Antimicrobial and Antioxidant Activities	Das et al. (2020)
	NiO nanoparticles	–	Spherical in shape	Anticancer activity	Ezhilarasi et al. (2016)
	ZnS nanoparticles	30 nm	Spherical in shape	Optical, dielectric and electronic	Sur and Ankamwar, (2016)
	TiO_2 nanoparticles	100 nm	Spherical morphology	Wound healing activity	Sivaranjani and Philominathan (2015)
	Gold nanoparticles	20–60	Spherical morphology	–	Anirban et al. (2013)

(continued)

Table 13.1 (continued)

Moringa Plant parts used for synthesis	Nanomaterials	Size	Shape	Application	References
Seed	Iron oxide nanoparticles	–	–	Antibacterial action in wastewater	de Oliveira et al. (2020)
	Magnetite nanoparticles	–	–	Recovery of Precious Metal Ions from Aqueous solution	Amuanyena et al. (2019)
	Silver nanoparticles	–	–	Anti-Methicillin resistant Staphylococcus aureus activity	Kalugendo and Kousalya (2019)
Stem Bark	Silver nanoparticles	40 nm	Spherical- and Pentagon-shaped particles	Anticancer activity	Vasanth et al. (2014)
	ZnO nanoparticles	–	–	–	Manokari and Shekhawat (2016)
Flower	ZnO nanoparticles	–	–	–	Manokari and Shekhawat (2016)
	Silver nanoparticles	8 nm	Monodispersed spherical nanoparticles	Antimicrobial and Sensing application	Bindhua et al. (2020)
Fruit pods	ZnO nanoparticles	–	–	–	Manokari and Shekhawat (2016)
Gum	Gold nanoparticles	20–80 nm	Individually dispersed irregular-shaped particles	antimicrobial activity	Ponnanikajamideen et al. (2016)
Gum	Nanogels	–	Gel with 8% of mucilage was effective	Drug delivery	Panda et al. (2006)

References

Ali J, Ali N, Wang L, Waseem H, Pan G (2019) Revisiting the mechanistic pathways for bacterial mediated synthesis of noble metal nanoparticles. J Microbiol Methods 159:18–25

Aliyu AO, Garba S, Bognet O (2017) Green synthesis, characterization and antimicrobial activity of vanadium nanoparticles using leaf extract of Moringa oleifera. Int J Chem Sci 16:231

Amuanyena MON, Kandawa-Schulz M, Kwaambwa HM (2019) Magnetic iron oxide nanoparticles modified with Moringa seed proteins for recovery of precious metal ions. J Biomater Nanobiotechnology 10:142–158

Anirban C, Dipesh D, Mahuya S, Sanjit D, Sekhar B (2013) Moringa oleifera leaf extract mediated green synthesis of stabilized gold nanoparticles. J Bionanosci 7:415–419

Ashokkumar T, Prabhu D, Geetha R, Govindaraju K, Manikandan R, Arulvasu C, Singaravelu G (2014) Apoptosis in liver cancer (HepG2) cells induced by functionalized gold nanoparticles. Colloids Surf B 123:549–556

Bindhua MR, Umadevi M, Esmail GA, Al-Dhabi NA, Arasu MV (2020) Green synthesis and characterization of silver nanoparticles from Moringa oleifera flower and assessment of antimicrobial and sensing properties. J Photochem Photobiol B Biol 205:111836

Das PEd, Majdalawieh AF, Abu-Yousef IA, Narasimhan S, Poltronieri P (2020) Use of a hydroalcoholic extract of Moringa oleifera leaves for the green synthesis of bismuth nanoparticles and evaluation of their anti-microbial and antioxidant activities. Materials 13:876

de Oliveira AM, Mateus GAP, dos Santos TRT, de Abreu Filho BA, Gomes RG, Bergamasco R (2020) Functionalized magnetite nanoparticles with Moringa oleifera with potent antibacterial action in wastewater. Environ Technol 1–10

Dorcheh S.K., Vahabi K. (2016) Biosynthesis of nanoparticles by fungi: large-scale production. In: Mérillon JM, Ramawat K (eds) Fungal metabolites. Reference series in phytochemistry. Springer, Cham

Eugenio M, Müller N, Frases S, Almeida-Paes R, Lima LMTR, Lemgruber L, Farina M, Souza de W, Sant Anna C (2016) Yeast-derived biosynthesis of silver/silver chloride nanoparticles and their antiproliferative activity against bacteria. RSC Adv 6:9893–9904

Ezhilarasi A, Judith Vijaya J, Kaviyarasu K, Maaza M, Ayeshamariam A, John Kenned L (2016) Green synthesis of NiO nanoparticles using Moringa oleifera extract and their biomedical applications: cytotoxicity effect of nanoparticles against HT-29 cancer cell, 164:352–360

Feng QL, Wu J, Chen GQ, Cui FZ, Kim TN, Kim JO (2000) A mechanistic study of the antibacterial effect of silver ions on Escherichia coli and Staphylococcus aureus. J Biomed Mater 52:662–668

Govindaraju K, Tamilselvan S (2020) Green synthesis of metal and metal oxide nanomaterials using seaweed bioresources. In: Rai M, Golinska P (eds) Microbial nanotechnology. CRC Press, New York pp 66–86

Govindaraju K, Kiruthiga V, Ganesh Kumar V, Singaravelu G (2009) Extracellular synthesis of silver nanoparticles by a marine alga, *Sargassum wightii* Greville and their antibacterial effects. J Nanosci Nanotechnol 9:5497–5501

Govindaraju K, Kiruthiga V, Manikandan R, Ashokkumar T, Singaravelu G (2011) β-glucosidase assisted biosynthesis of gold nanoparticles: a green chemistry approach. Mater Lett 65:256–259

Govindaraju K, Vasantharaja R, Uma Suganya KS, Anbarasu S, Revathy K, Pugazhendhi A, Karthickeyan D, Singaravelu G (2020a) Unveiling the anticancer and antimycobacterial potentials of bioengineered gold nanoparticles. Process Biochem 96:213–219

Govindaraju K, Vijai Anand K, Anbarasu S, Theerthagiri J, Revathy S, Krupakar P, Durai G, Kannan M, Subramanian KS (2020b). Seaweed (*Turbinaria ornata*)-assisted green synthesis of magnesium hydroxide [$Mg(OH)_2$] nanomaterials and their antimycobacterial activity. Mater Chem Phys 239:122007

Govindaraju K, Tamilselvan S, Kannan M, Kathickeyan D, Shkolnik D, Chaturvedi S (2020c) Nanomicronutrients [γ-Fe_2O_3 (iron) and ZnO (zinc)]: green preparation, characterization, agro-morphological characteristics and crop productivity studies in two crops (rice and maize). New J Chem 44:11373–11383

Guilger-Casagrande M, Lima R (2019) Synthesis of silver nanoparticles mediated by fungi: a review. Front Bioeng Biotechnol 7:287

Guo S, Wang E (2011) Noble metal nanomaterials: Controllable synthesis and application in fuel cells and analytical sensors. Nano Today 6:240–264

Hamouda RA, Hussein MH, Abo-elmagd RA, Bawazir SS (2019) Synthesis and biological characterization of silver nanoparticles derived from the cyanobacterium Oscillatoria limnetica. Sci Rep 9:13071

Iravani S (2014) Bacteria in nanoparticle synthesis: current status and future prospects. Int Sch Res Notices, 359316

Itroutwar PD, Govindaraju K, Tamilselvan S, Kannan M, Raja K, Subramanian KS (2020) Seaweed-based biogenic ZnO nanoparticles for improving agro-morphological characteristics of rice (Oryza sativa L.). J Plant Growth Regul 39:717–728

Kalugendo E, Kousalya P (2019) Synthesis of silver nanoparticles using *Moringa oleifera* seeds, *Glycyrrhiza glabra* stems, and its anti-methicillin-resistant *Staphylococcus aureus* activity. Asian J Pharm Clin Res 12:368–370

Katata-Seru L, Moremedi T, Aremu OS, Bahadur I (2018) Green synthesis of iron nanoparticles using Moringa oleifera extracts and their applications: removal of nitrate from water and antibacterial activity against Escherichia Coli, vol 256, pp 296–304

Khaleel Basha S, Govindaraju K, Manikandan R, Ahn JS, Bae EY, Singaravelu G (2010) Phytochemical mediated gold nanoparticles and their PTP 1B inhibitory activity. Colloids Surf B Biointerfaces 75:405–409

Madi N, Dany M, Abdoun S, Usta J (2016) Moringa oleifera's nutritious aqueous leaf extract has anticancerous effects by compromising mitochondrial viability in an ROS-dependent manner. J Am Coll Nutr 35:604–613

Manimaran M, Kannabiran K (2017) Actinomycetes-mediated biogenic synthesis of metal and metal oxide nanoparticles: progress and challenges. Lett Appl Microbiol 64:401–408

Manokari M, Shekhawat MS (2016) Zinc oxide nanoparticles synthesis from Moringa oleifera Lam. Extracts and their characterization. World Sci News 55:252–262

Moghaddam AB, Namvar F, Moniri M, Tahir PMd, Azizi S, Mohamad R (2015) Nanoparticles biosynthesized by fungi and yeast: a review of their preparation, properties, and medical applications. Molecules 20:16540–16565

Moodley JS, Krishna SBN, Sershen KP, Govender P (2018). Green synthesis of silver nanoparticles from Moringa oleifera leaf extracts and its antimicrobial potential. In: Advances in natural sciences: nanoscience and nanotechnology, vol 9, pp 015011

Mukherjee S, Nethi SK (2019) Biological synthesis of nanoparticles using bacteria, nanotechnology for agriculture, pp. 37–51

Narwade B, Prasad N, Lokhande SM, Madavi AB, Sahoo AK (2018) Extracellular biosynthesis of silver nanoparticles using Moringa oleifera leaves extract and its antimicrobial efficacy in packaging materials. Res J Life Sci Bioinform Pharm Chem Sci 4:188–202

Nilanjana G, Samrat P, Piyali B (2014). Silver nanoparticles of Moringa oleifera–green synthesis, characterisation and its antimicrobial efficacy. J Drug Deliv Ther 42–46

Pal S, Mondal S, Maity J, Mukherjee R (2018) Synthesis and characterization of ZnO nanoparticles using Moringa Oleifera leaf extract: Investigation of photocatalytic and antibacterial activity. Int J Nanosci Nanotechnol 14:111–119

Panaek A, Kvitek L, Prucek R, Kolar M, Veerova R, Pizurova N, Sharma VK, Nevena T, Zboril R (2006) Silver colloid nanoparticles: synthesis, characterization and their antibacterial activity. J Phys Chem B 110:16248–16253

Panda D, SI S, Swain S, Kanubgo SK, Gupta R (2006) Preparation and evaluation of gels from gum of Moringa oleifera. Indian J Pharm Sci 68(6):777–780

Ponnanikajamideen M, Rukumani V, Paulkumar K, Vanaja M, Samuel Rajendran R, Annadurai G (2016) Moringa oleifera Gum mediated fabrication of gold nanoparticles for effective DNA protection. J Bionanosci 10:1–7

Prabhu D, Arulvasu C, Babu G, Manikandan R, Srinivasan P, Govindaraju K, Ashokkumar T (2014) Synthesis and characterization of silver nanoparticles using crystal compound of sodium para-hydroxybenzoate tetrahydrate isolated from *Vitex negundo*. L leaves and its apoptotic effect on human colon cancer cell lines. Eur J Med Chem 84:90–99

Prasad TNVKV, Elumalai EK (2011) Biofabrication of Ag nanoparticles using *Moringa oleifera* leaf extract and their antimicrobial activity. Asian Pac J Trop Biomed 1:439–442

Sandbhy V, MacBride MM, Peterson BR, Sen A (2006) Silver bromide nanoparticles/polymer composites: dual action tunable antimicrobial materials. J Am Chem Soc 128:9798–9808

Sathyavathi R, Bala Murali Krishna M, Narayana Rao D (2010) Biosynthesis of silver nanoparticles using Moringa oleifera leaf extract and its application to optical limiting. J Nanosci Nanotechnol 10:1–5

Setua P, Chakraborty A, Seth D, Bhatta MU, Satyam PV, Sarkar N (2007) Synthesis, optical properties, and surface enhanced Raman scattering of silver nanoparticles in nonaqueous methanol reverse micelles. J Phys Chem C 111:3901–3907

Shousha WG, Aboulthana WM, Salama AH, Saleh MH, Essawy EA (2019) Evaluation of the biological activity of Moringa oleifera leaves extract after incorporating silver nanoparticles, in vitro study. Bull Natl Res Centre 43:212

Singaravelu G, Arockiayamari JS, Ganesh Kumar V, Govindaraju K (2007) A novel extracellular synthesis of monodisperse gold nanoparticles using marine algae, *Sargassum wightii*. Colloids Surf B 57:97–101

Sinha S, Pan I, Chanda P, Sen SK (2009) Nanoparticles fabrication using ambient biological resources. J Appl Biosci 19:1113–1130

Sivaranjani V, Philominathan P (2015) Synthesize of Titanium dioxide nanoparticles using Moringa oleifera leaves and evaluation of wound healing activity. 12:1–5

Stobie N, Duffy B, McCormack DE, Colreavy J, Hidalgo M, McHale P (2008) Prevention of Staphylococcus epidermisdis biofilm formation using a low temperature processed silver doped phenyltriethoxysilane solgel coating. Biomater 29:963–969

Sura UK, Ankamwar B (2016) Optical, dielectric, electronic and morphological study of biologically synthesized zinc sulphide nanoparticles using Moringa oleifera leaf extracts and quantitative analysis of chemical components present in the leaf extract. RSC Adv 6:95611–95619

Thangavelu RM, Ganapathy R, Ramasamy P, Krishnan K (2020) Fabrication of virus metal hybrid nanomaterials: an ideal reference for bio semiconductor. Arab J Chem 13:2750–2765

Tiloke C, Anand K, Gengan RM, Chuturgoon AA (2018) Moringa oleifera and their phytonanoparticles: potential antiproliferative agents against cancer. Biomed Pharmacother 108:457–466

Uma Suganya KS, Govindaraju K, Ganesh Kumar V, Stalin Dhas T, Karthick V, Singaravelu G, Elanchezhiyan M (2015a) Size controlled biogenic silver nanoparticles as antibacterial agent against isolates from HIV infected patients. Spectrochim Acta Part A Mol Biomol Spectrosc 144:266–272

Uma Suganya KS, Govindaraju K, Ganesh Kumar V, Stalin Dhas T, Karthick V, Singaravelu G, Elanchezhiyan M (2015b) Blue green alga mediated synthesis of gold nanoparticles and its antibacterial efficacy against Gram positive organisms. Mater Sci Eng C 47:351–356

Uma Suganya KS, Govindaraju K, Ganesh Kumar V, Prabhu D, Arulvasu C, Stalin Dhas T, Karthick V, Changmai N (2016a) Anti-proliferative effect of biogenic gold nanoparticles against breast cancer cell lines (MDA-MB-231 & MCF-7). Appl Surf Sci 371:415–424

Uma Suganya KS, Govindaraju K, Ganesh Kumar V, Karthick V, Krupakar P (2016b) Pectin mediated gold nanoparticles induces apoptosis in mammary adenocarcinoma cell lines. Int J Biol Macromol 93:1030–1040

Vasanth K, Ilango K, MohanKumar R, Agrawal A, Dubey GP (2014) Anticancer activity of Moringa oleifera mediated silver nanoparticles on human cervical carcinoma cells by apoptosis induction. Colloids Surf, B 117:354–359

Venkatachalam M, Govindaraju K, Mohamed Sadiq A, Tamilselvan S, Ganesh Kumar V, Singaravelu G (2013) Functionalization of gold nanoparticles as antidiabetic nanomaterial. Spectrochim Acta Part A Mol Biomol Spectrosc 116:331–336

14 Moringa Processing and Value Addition: Rice Analogue From Broken Rice Fortified With Moringa Leaves and Its Effect on Iron Availability

S. Amutha, R. Yogeshwari, and G. Hemalatha

Abstract

Attempts were made in terms of formulating product ingredients, optimizing extrusion conditions, evaluating quality characteristics for developing fortified extruded rice analogues. In order to develop fortified rice, broken rice flour was selected as a base carrier, and synthetic fortificants were screened based on bioavailability and sensory acceptability viz., micronized ferric pyrophosphate (MFPP) as iron source, and natural fortificants were screened based on high iron content from plant foods viz., moringa leaf powder as iron source. Formulations of chemical fortificants that should be added into the broken rice flour were calculated (to meet 33% RDA of adult woman iron requirement) as 7 mg for the target groups. In addition to the RDA, excess amount of micronutrients was added to compensate for the losses during extrusion and cooking process. Standardization was carried out by performing sensory evaluation to determine the incorporation levels such as 10% moringa leaf powder for development of naturally fortified extruded rice analogues fortified as iron source. Based on the interactions of micronutrients, five treatments were standardized. As per the standardized treatments, fortificants were dry blended with broken rice flour except T_0 (commercial rice) and T_1 (unfortified extruded rice). Thus, chemically fortified extruded rice analogues contained 7.65 mg of iron. Fortified extruded rice analogues were analysed for physical, chemical, cooking and sensory characteristics. The overall retention of iron in cooked rice analogues was 100% for iron in both chemical and natural fortificants.

14.1 Introduction

Moringa (*Moringa oleifera)* is the most popular, perennial vegetable grown in south India. The tree is commonly grown for its unique vegetable. Its fruits, leaves and flowers are also equally useful. The drumstick tree is a softwood tree and is native to India. The trees are grown in the tropical and sub-tropical regions around the world. For long term use and storage, moringa leaves may be dried and powdered to preserve their nutrients. Sun, shade, freeze and oven drying at 50–60 °C are all acceptable methods; however, they are variable in their retention efficacy of specific micro and macronutrients. The powder is commonly added to soups, sauces and smoothies. Owing to its high nutritional density (see Chap. 3), moringa leaf powder is valued as a dietary supplement and may be used

S. Amutha (✉) · R. Yogeshwari · G. Hemalatha
Community Science College and Research Institute, Tamil Nadu Agricultural University, Madurai, India
e-mail: deanhscmdu@tnau.ac.in

N. M. Boopathi et al. (eds.), *The Moringa Genome*, Compendium of Plant Genomes,
https://doi.org/10.1007/978-3-030-80956-0_14

to enrich food products with acceptable sensory evaluation.

Since rice is an important staple for more than one-third of the population (and more particularly, throughout India), it is obviously an excellent vehicle for delivering micronutrients to large number of people and has the potential to significantly alleviating micronutrient deficiencies. In addition, fortification of rice will allow its consumers to benefit without making major changes in their dietary habits (Kunz 2009).

Broken rice has low economic value as compared to whole rice and are not well accepted by consumers, although broken rice is nutritionally as good as whole rice itself. Generally, broken rice is of poor quality due to admixture with grit, stones and clay particles. Extrusion technology is a promising technology to improve the quality of broken rice. Extrusion technology is a versatile technology that utilizes a single screw or a set of screws to force food materials through a small opening. While food is being forced through the extruder, foods are cooked by the high pressure, high shear and high temperature environment created by the screws, which are encased in the barrel (Fig. 14.1).

Hence, extrusion processing gives the option of reforming the broken rice flour into a pre-cooked product that can be shaped and have a texture similar to that of cooked rice kernel. In this process, different vitamins, minerals, antioxidants and other potential functional ingredients can be added inside the body of rice. The rice developed by extrusion technology does not need to be soaked in water for a long time prior to cooking, and therefore, all the added ingredients will be retained within the rice.

Extrusion technology is increasingly being employed in the production of extruded analogues of various shapes and sizes from rice flour. They can be made from formulations comprising of broken rice and other non-rice ingredients which are added to improve the quality of the extrudates (Mishra et al. 2012).

14.2 Iron Deficiency

Iron is an important micronutrient which is essential for various functions in human body. Iron Deficiency Anaemia (IDA) affects population of all age groups, but adolescent girls are more vulnerable to anaemia and its associated complications. The requirement for iron, in fact, doubles during adolescence as compared to younger age. There is a significant increase in the requirement of iron from preadolescent stage of approximately 0.7–0.9 mg iron per day to as much as 1.37–1.88 mg per day in adolescent boys and 1.40–3.27 mg per day in adolescent girls (Tesfaye et al. 2015). As prevalence of anaemia is 50%, it needs strategic intervention for its prevention and control (Kumari et al. 2017). Hence, it is envisaged that in order to alleviate micronutrient malnutrition among rice-eating population in the world, micronutrients like iron-fortified rice analogues, if made using broken rice flour by extrusion technology, would be of great significance to the health and nutrition of the present generation. Hence, this chapter

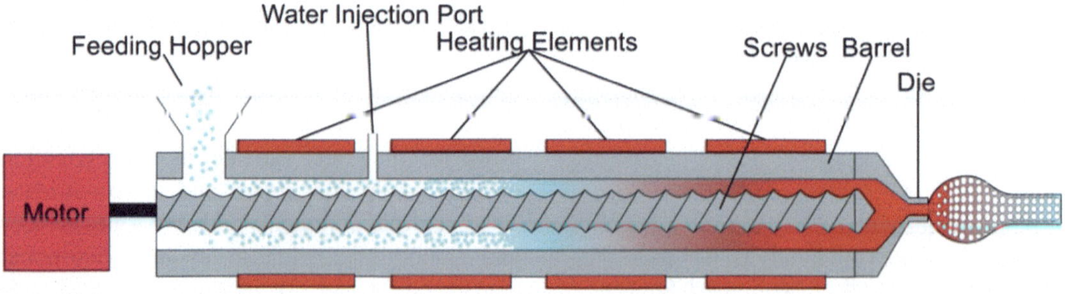

Fig. 14.1 A simple schematic of extrusion processing, showing the transformation of raw ingredients to finished product

focuses on rice analogue from broken rice fortified with moringa leaves and its effect on iron.

14.3 Preparation of Broken Rice Flour

The broken rice of variety IR 20 was procured from Modern rice mill, and it was cleaned to remove dirt and discoloured grains and then washed with tap water and sun-dried. They were made into flour using a burr mill which were then sieved through BS 60 metal laboratory sieve to obtain homogenous particle size and stored in airtight containers.

14.4 Screening of Synthetic Fortificants for Developing Fortified Extruded Rice

Technically, fortification is one of the most challenging techniques of increasing the nutrient content of foods, because the fortificants that have the best bioavailability tend to be those that interact most with food constituents to produce undesirable organoleptic changes. When selecting a suitable micronutrient as a fortificant, the overall objective is to find the one that has the greatest absorbability, *i.e.* the highest relative bioavailability and at the same time which does not cause unacceptable changes to the sensory properties (*i.e.* taste, colour, texture) of the food vehicle.

Cost is usually another important consideration. For fortification of rice analogues with iron, Planning and Appropriate Technology for Health (2008) recommended the use of Micronized Ferric Pyrophosphate (MFPP) compounds having 24–26% iron, with an average particle size of 3 microns. The MFPP is manufactured with small particle size (and thus high surface area) such that it has better bioavailability than most non-soluble forms of iron.

14.5 Formulation of Micronutrient Fortification Level for Development of Fortified Extruded Rice Analogues

The main aim of formulating the level of fortificants (Table 14.1) in fortified extruded rice analogues was to preserve the nutritional balance and safety of the diet for the target population.

Formulations of fortificants that should be added into the broken rice flour was calculated to meet 33% RDA of adult woman for iron 7 mg, and vitamin C 13.3 mg/100 g for the target population, in addition to the RDA, excess amount of micronutrients was added to compensate the expected losses during extrusion processing, cooking and storage. Hence, the final target amount of fortification was computed as 6.83 mg of iron and 36.57 mg of vitamin C per 100 g of rice. Based on the interactions of micronutrients, the different treatments (T_0-T_5) were standardized as controls (T_0-Control-Commercial rice (IR 20), T_1-Unfortified Extruded rice-Without fortification), synthetic fortification (T_2-MFPP-Broken rice flour fortified with Micronized Ferric Pyrophosphate (MFPP), T_3-AA-Broken rice flour fortified with Ascorbic Acid) and natural fortification (T_4-MLP-Broken rice flour fortified with Moringa Leaf Powder, T_5-L-Broken rice flour fortified with Lemon juice).

14.6 Extruded Rice Analogue Development

Limited studies are available for development of rice analogues that employed single screw extruder. Hence, approximate level of moisture content and additive levels were obtained from the literature that focussed on development of extruded rice analogues (Hussain et al. 2017). Further, to find out the appropriate levels of moisture and additive levels, Response Surface

Table 14.1 Suggested fortificant levels for fortified rice at moment of consumption

Nutrient	Micronutrient compound	Rice intake/day	Fortificant concentration /100 g of rice
Iron	Micronized ferric pyrophosphate	75–149 g	12 mg
Vitamin C	Ascorbic acid		26 mg

Methodology (RSM) was used based on variables and actual levels. RSM is a statistical and mathematical tool that allows the evaluation of the effect of multiple variables and their interactions on the output variables with reduced number of trials, and thus, it has a major advantage over the one-factor-at-a-time approach.

The extruded rice analogues were processed by the following systematic procedures as advocated for the single screw extruder machine. The product ingredients such as broken rice flour, distilled mono glyceride and water were mixed, and additionally, salt was added at a level of 2% by weight to all treatments. This was followed by dry mixing, addition of water and kneading for 30 min in upper mixer tank attached in extruder to obtain dough. Then, the dough was extruded, with the cutter speed which was set up at 1000 rpm/min at the point of extrusion. The temperature of the extrudates coming out from the rice-shaped die was measured to be 62°C, and hence, it was noted that the final product was partially cooked. The extruded rice analogues were further steamed for 3 min, cooled and dried under shade drying for overnight to reach a moisture content of 11% (Fig. 14.2).

14.7 Determination of Micronutrient Composition of Rice Flour (Before and After Extrusion)

Formulated target amount of chemical fortificants, namely, micronized ferric pyrophosphate (MFPP) and ascorbic acid were mixed with rice flour by using mixer, attached in extruder machine, for 30 min. Micronutrient compositions including iron were determined in unfortified rice flour and rice flour fortified with micronized ferric pyrophosphate MFPP (iron source) and ascorbic acid (Vitamin C source). The procedure for Iron and Vitamin C estimation was adopted as per the method described in AOAC–Association of Official Agricultural Chemists and Horwitz (1975).

After determining micronutrient composition in chemically fortified rice flour, it was utilized to develop dried rice analogues and analysed to determine the retention of fortified micronutrients as described above.

Plant-based foods such as moringa leaf powder (iron source) and lemon juice (vitamin C source) were chosen as source of natural fortificants. The proximate and micronutrient composition such as moisture, protein, fat, fibre, ash, carbohydrate, ascorbic acid and iron of natural fortificants was estimated as detailed above.

14.8 Standardization of Natural Fortificant Levels

The addition of natural fortificants such as moringa leaf powder and lemon juice (Fig. 14.3) may influence the sensory properties of the rice analogues. Sensory characteristics play an important role in consumer acceptability apart from nutritional qualities. Therefore, standardization was carried out by performing sensory evaluation with incorporation of each natural fortificants at different levels into broken rice flour to obtain better sensory qualities of the fortified extruded rice analogues. The sensory score card used in this study was in 9 to 1 Point hedonic rating scale.

For developing naturally fortified extruded rice analogues, micronutrients such as iron and ascorbic acid were estimated in broken rice flour fortified with moringa leaf powder (T_4) and lemon juice (T_5) by processing moringa and

Fig. 14.2 Schematic illustration on development of fortified extruded rice analogues

lemon as illustrated in Fig. 14.3. After determining micronutrient composition of iron and ascorbic acid in naturally fortified rice flour, it was utilized to develop dried rice analogues and analysed to determine the retention of micronutrients.

14.9 Quality Evaluation of Fortified Extruded Rice Analogues

Physical characteristics of the fortified extruded rice analogues (T_0 – T_5) such as moisture, thousand grain weight, grain length, grain breath, bulk density, true density, porosity and sensory score were evaluated and compared with natural rice. The detailed procedure adopted to evaluate the physical properties was as per the method described by AOAC–Association of Official Agricultural Chemists and Horwitz (1975). The chemical constituents of fortified extruded rice analogues were also analysed for protein, fat, crude fibre, starch, amylose and amylopectin contents using different procedures as outlined in Table 14.2. The procedure followed was given below.

14.9.1 Cooking Characteristics of the Fortified Extruded Rice Analogues

Cooking properties were evaluated for fortified extruded rice analogues. A boiling tube (50 ml) containing rice analogues (2 g) and 20 ml distilled water was placed in boiling water bath at 98 °C. It was cooked until white core disappeared at the centre of 90% of the rice analogues, i.e., minimum cooking time. The cooked sample was then weighed, on completion of cooking. The cooking time was calculated as the minimum cooking time Juliano et al. (1981). Water uptake ratio was calculated using the following formula described by Juliano et al. (1982). Rice analogue was cooked for optimum cooking time. The volume expansion ratio was estimated using the following formulae described by Juliano (1985a, b). The length of rice analogues was measured

Fig. 14.3 Preparation of Moringa leaf powder and lemon juice for natural fortification

Table 14.2 Procedures used to estimate different chemical constituents of rice analogue

S.no	Chemical constituents	Methods/Instruments	Reference
1	Protein	Micro Kjeldahl	Ma and Zuazaga (1942)
2	Fat	Soxhlet apparatus	Cohen et al. (1986)
3	Crude fibre	Fibra plus	Zebeli et al. (2012)
5	Amylose	Iodometry method	Sadasivam and Manickam (1992)
6	Amylopectin	Iodometry method	Sadasivam and Manickam (1992)

using a digital vernier calliper of ±0.01 mm accuracy and a graph paper (Juliano 1971). The drained cooking water was evaporated to dry in a tared evaporating dish to evaporate the water as steam in a water bath at 80 °C and then evaporated to dryness in a hot air oven at 105 °C (8 h), and the weight of the residue is noted to determine the loss of solids (Juliano et al. 1982). From the fortified extruded rice analogues, different foods viz., tamarind rice, coconut rice, pepper rice and mint rice have been developed and evaluated for consumer acceptability and were statistically analysed as described by Snedecor and Cochran (1994).

Naturally prevailing iron content in rice flour was found to be 0.82 mg per 100 g. In order to obtain required amount of micronutrients for daily human need, chemical fortificants of micronized ferric pyrophosphate (6.83 mg) and ascorbic acid (36.57 mg) were mixed into 100 g of broken rice flour as mentioned in Table 14.1. Chemical fortificants are added in excess until the concentration of iron and vitamin C had reached above the required levels, to compensate the loss that was expected during extrusion, cooking and storage.

After determining micronutrient composition in chemically fortified rice flour, it was utilized to develop dried rice analogues and analysed to determine the retention of fortified micronutrients. Chemically fortified extruded rice analogues contained 7.65 mg of iron with 100% retention and vitamin C content was found to be 26.33 mg/100 with 71.99% retention. Thus, it was observed that no iron was lost during extrusion processing, as iron do not degrade as easily as the vitamins.

On the other hand, the retention of vitamins in extruded rice was decreased. This may be due to extrusion conditions such as increased temperature, increased screw speed, decreased moisture, decreased throughput and decreased die diameter.

The proximate and micronutrient composition (moisture, protein, fat, fibre, ash, carbohydrate, ascorbic acid and iron) of natural fortificants was estimated (Table 14.3) and graphically represented in Fig. 14.4.

14.9.2 Standardization of Natural Fortificants Levels for Developing Extruded Rice Analogues

Moringa leaf powder fortification level was determined by optimizing the concentration. The sensory scores of organoleptic evaluations for moringa leaf powder fortified rice analogues was carried out for optimizing the concentration of moringa leaf powder into broken rice flour with different incorporation level (6, 8, 10, and 12%).

Based on the sensory evaluation, the sensory scores for all parameters were increased up to 10% incorporation level of moringa leaves powder and also overall acceptance score was high in 10% incorporation level. The control formulation (0% moringa leaf powder) had the lowest overall acceptability scores of 4.1–4.4. Hence, addition of 10% moringa leaf powder was selected for the development of fortified extruded rice analogues (Fig. 14.5) and it is used for further study. Similar results have been found by Liu et al. (2011) who reported that the sensory scores of the developed extruded snacks based on moringa leaf powder and oat flour had highest overall acceptability scores (6.9–7.5) at 30 and 45% addition of moringa leaf powder. As per the sensory evaluation, incorporation levels at 10% moringa leaf powder and 30% lemon juice were standardized for development of naturally fortified extruded rice analogues for fortification of iron and ascorbic acid, respectively.

Micronutrients such as iron and ascorbic acid were estimated in broken rice flour fortified with 10% moringa leaf powder (T_4) and 50% lemon juice (T_5). After mixing of standardized natural fortificant levels into broken rice flour, the concentrations of iron and ascorbic acid were found to have 1.48 mg and 20.21 mg/100 g in T_4 and T_5, respectively. These values are below the level of target amount of fortification because addition of natural fortificants was optimized based on sensory test.

Moisture content of the control natural rice (variety IR 20) was found to be 10.58%, whereas the the fortified extruded rice analogues had 10.47–10.76%, and hence, there is no significant

Table 14.3 The proximate and micronutrient composition of natural fortificants

Natural fortificants	Moisture %	Protein %	Fat %	Ash %	Fibre %	Carbohydrate %	Ascorbic acid (mg/100 g)	Iron (mg/100 g)
Moringa leaf powder	7.43	18.49	5.13	7.15	9.25	45.52	0.72	7.42
Lemon juice	83.63	0.97	0.33	0.41	0.20	8.42	40.42	0.85

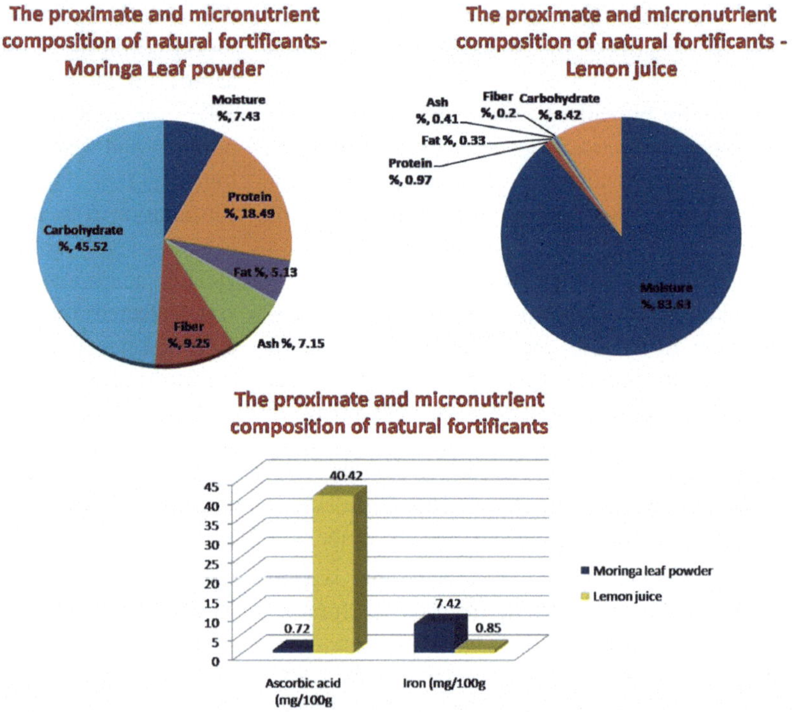

Fig. 14.4 The proximate and micronutrient composition of natural fortificants in Moringa leaf powder and lemon juice

difference in terms of moisture content ($p < 0.01$) among the control and extruded rice samples.

Significant difference in data for 1000 grain weight was noticed: natural rice had 17.09 g of 1000 grain weight, whereas the extruded rice samples (T_1–T_5) had range between 19.60 and 19.66 g which were higher than the natural rice (Table 14.4).

The grain length of the natural rice (T_0) was found to be 5.69 mm and the extruded rice analogues (T_1–T_5) had the range between 6.20 and 6.25 mm. The maximum length of 6.25 mm was found in Ascorbic acid (T_3)-fortified rice analogues and minimum length of 6.20 was found in unfortified extruded rice (T_1) (The statistical analysis revealed that significant difference ($p < 0.01$) was found between the control rice and extruded rice samples).

The grain breath of the natural rice (T_0) was 1.70 mm. After extrusion process, it was decreased in extruded rice analogues (T_1 – T_5) and ranged from 2.22 to 2.26 mm. Whereas maximum breath of 2.26 mm was found in unfortified extruded rice (T_1). Statistical analysis revealed the significant difference ($p < 0.01$) in

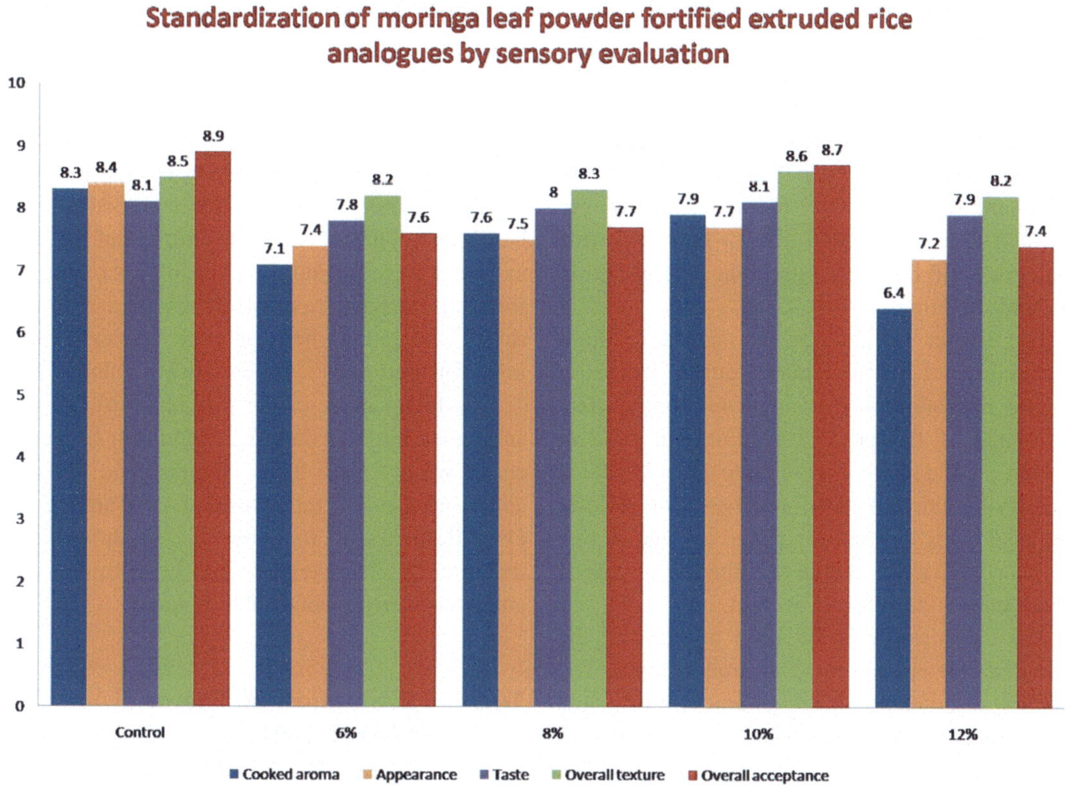

Fig. 14.5 Sensory evaluation score of extruded rice analogues fortified with different concentrations of Moringa leaf powder

Table 14.4 Physical characteristics and sensory scores of the extruded fortified rice analogues

Treatments	Moisture (%)	1000 grain weight (g)	Grain length (mm)	Grain breath (mm)	Bulk density (g/ml)	True density (g/ml)	Porosity (%)	Sensory score (OAA)
T_0	10.58	17.09^a	5.69^a	1.70^a	0.93^a	0.23^a	2.86^c	9.2^a
T_1	10.47	19.62^b	6.20^b	2.26^b	0.81^b	0.16^c	4.06^{ab}	8.7^b
T_2	10.72	19.60^b	6.24^b	2.24^b	0.81^b	0.15^d	4.10^a	8.6^{bc}
T_3	10.58	19.62^b	6.25^b	2.23^b	0.81^b	0.16^c	4.06^{ab}	8.5^{bc}
T_4	10.58	19.60^b	6.24^b	2.22^b	0.84^b	0.16^c	4.00^{ab}	8.5^{bc}
T_5	10.53	19.66^b	6.21^b	2.23^b	0.81^b	0.16^c	4.06^{ab}	8.3^c

^{abc}Same letter in the same column are on par and different letter in the same column are significantly different ($p \leq 0.01$)

breath values among the natural rice and extruded rice analogues. Hussain (2012) reported that higher value was obtained for the length and breadth of extruded rice analogues than natural rice which may be due to expansion during the extrusion process.

The bulk density is the ratio of mass of the rice analogues to its total (bulk) volume. Based on the analysis, extruded rice analogues had a lower density (0.81–0.84 g/ml) than natural rice (0.93 g/ml). The statistical analysis revealed significant difference ($p \leq 0.01$) between the bulk density values of control and extruded rice samples, wherein T_0 had higher bulk density value and T_1– T_5 were found to be on par between the treatments. Thus, extruded rice analogues had lighter weight than natural rice at the same volume.

The lighter weight can be caused by high temperature of extruder barrel during extrusion process and addition of water which may affect the shape of the product resulting in product swelling. Widara and Budijanto (2012) stated that drying process while practised for making rice analogues will reduce water content so that the matrix of rice analogue becomes more porous which causes decrease in weight.

The true density is the ratio of mass of the extruded rice analogues to its true volume. The value of true density of the natural rice (T_0) was 0.23 g/ml and that of extruded rice analogues (T_1–T_5) ranged between 0.15 and 0.16 g/ml. The statistical analysis revealed significant difference ($p \leq 0.01$) between the treatments, wherein T_0 was having higher true density and T_2 was having lower true density and they were found to be on par. And also, on par values were found for T_1, T_3, T_4, and T_2 between each group of treatments.

The porosity of the rice analogues is the ratio of the volume of internal pores in between the rice analogues to its bulk volume. From the results, porosity of the natural rice (T_0) was lower as 2.86% than the extruded rice analogues of 4.06 to 4.10% (T_1–T_5). Based on statistical analysis, significant difference ($p \leq 0.01$) was found between natural rice and extruded rice analogues. It is also revealed that lower porosity was found in T_0 and higher porosity was found in T_1–T_5 and were found to be on par between the treatments.

This result is in accordance with Pandiselvam and Thirupathi (2014) who observed that the porosity of rice decreased from 46.82 to 38.27% with increased bulk and true densities. They found that as moisture in grain increases, the rice volume also increases, and hence, the number of grains in a fixed volume decreases.

The data for overall acceptability (OAA) sensory score of natural rice (T_0) was 9.2 and the extruded rice analogues (T_1–T_5) had OAA of 8.3–8.7 which were lower than the natural rice. Except natural rice, unfortified extruded rice T_1 had higher sensory score (8.7) than the fortified extruded rice T_2 and T_3 (8.5–8.6). Among the fortified extruded rice, MFPP-fortified extruded rice T_2 scored higher value of 8.6 and lemon juice-fortified rice (T_5) scored lower value of 8.3.

The lower score of the fortified extruded rice analogues may be due to the addition of fortificants which may have changed the sensory characteristics of the extruded rice analogues. Overall, the results for sensory score showed that the sensory characteristics of the extruded rice analogue were in acceptable range.

14.9.3 Evaluation of Chemical Characteristics of Fortified Extruded Rice Analogues

Existence of significant difference between protein content of unextruded and extruded rice analogues indicate that was a marginal loss of protein during extrusion process (Table 14.5). The extrusion processing conditions such as temperatures and moistures and the interaction between proteins and carbohydrates, lipids and other chemical products may lead to mallard reaction (Yogeshwari et al. 2019) which is evident for the marginal lower level of protein for extruded rice. Several researchers have reported decreased protein content of extruded food products. Jiddere and Filli (2015) also reported decrease in protein content from 12.50 to 11.00/100 g with increase in extrusion temperature (from 100 to 130 °C).

The fat content of the natural rice (T_0) was 0.41 mg/100 g. After extrusion process, the fat content was decreased. Moringa leaf powder fortified extruded rice analogues (T_4) contain 0.79/100 g. The higher fibre content (1.26/100 g) was observed in lemon juice-fortified extruded

Table 14.5 Chemical characteristics of the extruded fortified rice analogues

Treatments	Protein (g/100 g)	Fat (g/100 g)	Fibre (g/100 g)	Ash (%)	Amylose (%)	Amylopectin (%)
T_0	5.05^{b}	0.41^{d}	0.95^{e}	0.89^{f}	18.65^{a}	59.09
T_1	3.48^{d}	0.23^{e}	0.62^{c}	1.35^{e}	16.76^{b}	58.00
T_2	3.47^{d}	0.22^{e}	0.63^{c}	1.34^{e}	16.87^{b}	58.14
T_3	3.47^{d}	0.23^{e}	0.63^{c}	1.34^{e}	16.98^{b}	58.14
T_4	5.19^{b}	0.79^{b}	0.95^{ec}	2.01^{c}	16.73^{b}	58.18
T_5	3.14^{e}	0.42^{d}	1.26^{d}	1.59^{d}	16.34^{b}	58.11
SED	0.0869	0.0096	0.0233	0.0335	0.2536	0.9083
CD (0.01)	0.2656**	0.0294**	0.0710**	0.1023**	0.7747**	2.7746 NS

^{abc}Same letter in the same column are on par and different letter in the same column are significantly different ($p \leq 0.01$)
**Highly significant, *Significant, NS—Non Significant

rice analogues and the lower protein content was observed in natural rice T_1 (0.62/100 g). Maximum amylose content of 16.98% was found in ascorbic acid-fortified extruded rice analogues T_3, and minimum amylose content of 16.87% was found in MFPP-fortified extruded rice analogues T_1. Natural rice had higher amylose content.

Concentration of amylose and amylopectin ratio is crucial regarding rice-based extrudates. Juliano (1985a, b) categorized rice flour according to presence of amylose concentration. This study has confirmed the use of rice variety IR 20 for developing rice analogues with better firmness as it contained 18.65% amylose and 59.09% amylopectin.

14.9.4 Evaluation of Cooking Characteristics of Fortified Extruded Rice Analogues

The cooking characteristics such as cooking time, water uptake, volume expansion, elongation ratio and cooking loss were estimated for natural rice (T_0), extruded unfortified rice analogues (T_1) and chemical fortified extruded rice analogues containing micronized ferric pyrophosphate (MFPP) T_2, ascorbic acid (AA) T_3 and naturally fortified extruded rice analogues containing moringa leaf powder (T_4) and lemon juice (T_5).

Compared with natural rice (T_0), which takes 17 min of boiling for complete cooking, the extruded rice analogues (T_1–T_5) required boiling in water for 5 min in order to prepare them for consumption (Table 14.6). Statistical analysis revealed significant difference ($p \leq 0.01$) between natural rice and extruded rice samples. Khairunnisa et al. (2017) reported that the cooking time of the rice analogues was only 2 min and 50 s, which is considered shorter than the average cooking time of polished rice (which takes 11 min).

Water uptake for the extruded rice analogues (T_1 to T_5) ranged from 184 to 186% which was much lower than the uptake reported for natural rice T_0 (296%) at optimum cooking time.

The volume expansion ratio was also determined as volume of cooked rice to volume of uncooked rice. The volume expansion ratio of natural rice (T_0) was higher (4.26) than the extruded rice analogues (T_1 to T_5) which ranged between 3.61 and 3.66. Similar trend was observed by Hussain (2012) who reported volume expansion ratio of the extruded rice premix was significantly lower (3.6) than natural rice (4.0). Elongation ratio of the rice samples was calculated by ratio of length of the cooked and uncooked rice. The elongation ratio was lower

Table 14.6 Cooking characteristics of fortified extruded rice analogues

Treatments	Cooking time	Water uptake (ml)	Volume expansion ratio	Elongation ratio	Cooking loss (%)
T_0	17	296	4.26	1.49	2.7
T_1	5	186	3.65	1.25	5.3
T_2	5	184	3.65	1.24	5.3
T_3	5	186	3.63	1.27	5.4
T_4	5	185	3.61	1.25	5.3
T_5	5	186	3.66	1.23	5.3
SED	0.1164	3.4322	0.0616	0.0331	0.1349
CD (0.01)	0.3557**	10.4843**	0.1880**	0.1011**	0.4119**

**Highly significant

(1.23–1.27) in extruded rice than that of natural rice (1.49).

Cooking loss is an important quality parameter. High cooking loss is undesirable as it indicates high solubility of starch that resulted into turbid cooking water, whereas low cooking loss resulted in the stickiness of the product, and thus, affects the texture of the rice food. Cooking loss was significantly higher among the extruded rice analogues and ranged from 5.2 to 5.5% while it was low in natural rice T_0 (2.7%). The high solid content in the cooking water indicated a greater cooking loss due to leaching of gelatinized starch particles into the cooking water.

It was observed that iron contents of fortified rice flour (before extrusion) and fortified extruded rice analogues (rice after extrusion, drying and cooking) were relatively similar, as expected, because iron is not sensitive to heat, light and oxygen. However, only negligible decrease in iron was observed in naturally fortified extruded rice analogues. This observed loss of iron could be attributed to a settling of minerals in the extruder, causing extruded samples to contain a lower concentration of minerals. In other words, the flour sample was a homogenous mixture of rice flour, vitamins and minerals, while the rice analogue samples contained a greater portion of rice flour.

Generally, the retention of vitamins in extruded products decreases with several extrusion conditions namely increased temperature, screw speed, decreased moisture, decreased die diameter and increased specific energy input. Vitamin C is reported as being the least stable vitamin compared to other common vitamins used in enrichment and fortification of cereals.

Thus, it can be concluded that Moringa processing significantly add values to the rice products as.

- The extruded rice analogues had 100% retention of iron from chemical and natural fortificants.
- The extruded rice analogues had short cooking time (5 min), and hence, the technology can be taken advantage of, in promoting the fortified extruded rice as a convenient ready-to-use food.
- The versatility of the technique of quick cooking fortified extruded rice with desired nutrition profile can be tailor made to suit the specific nutrition needs of the malnourished population.
- The technology is flexible, making possible wide variations in product ingredients like addition of pulses, vegetables to increase the nutrient density of the product.
- Micronutrient-fortified rice can be a cost-effective and sustainable solution to reduce the burden of micronutrient deficiency in developing countries. Moringa leaf, dried and incorporated at 10% level, improves the iron content of fortified extruded rice analogues right from processing till the end of final product. It can be considered as a readily

available iron source, and thus can be a potential candidate for nutraceutical industries.

References

AOAC–Association of Official Agricultural Chemists, & Horwitz, W. , 1975.AOAC–Association of Official Agricultural Chemists, Horwitz W (1975) Official methods of analysis, vol 222. Association of Official Analytical Chemists, Washington, DC

Cohen LA, Thompson DO, Choi K, Karmali RA, Rose DP (1986) Dietary fat and mammary cancer. II. Modulation of serum and tumor lipid composition and tumor prostaglandins by different dietary fats: association with tumor incidence patterns. J Natl Cancer Inst 77(1):43–51

Hussain SZ (2012) Development of technology for the production of rice-premix fortified with micronutrients by extrusion process. PAU Ludhiana

Hussain SZ, Afshana B, Amin T (2017) Utilization of broken rice and walnut kernels for development of nutritious snacks using extrusion technology. Pharma Innov J 6(10):91–101

Jiddere G, Filli KB (2015) The effect of feed moisture and barrel temperature on the essential amino acids profile of sorghum malt and bambara groundnut based extrudates. J Food Process Technol 6(5):1

Juliano B (1971) A simplified assay for milled rice amylose. Cereal Sci Today 16:334–360

Juliano B (1985a) Polysaccharides, proteins, and lipids of rice. Rice Chem Technol 59–174

Juliano B (1985b) The rice grain and its gross composition. Rice Chem Technol

Juliano B, Perez CM, Blakeney AB, Castillo T, Kongseree N, Laignelet B, Webb BD (1981) International cooperative testing on the amylose content of milled rice. Starch/Stlrke 33(5)

Juliano B et al (1982) International cooperative testing of the alkali digestibility values for milled rice. Starch-Stärke 34(1):21–26

Khairunnisa K et al (2017) Formulation of high protein rice analog made pf cassav, maize starch and soybean. Int Joint Semin Symp

Kumari R, Bharti RK, Singh K, Sinha A, Kumar S, Saran A, Kumar U (2017) Prevalence of iron deficiency and iron deficiency anaemia in adolescent girls in a tertiary care hospital. J Clin Diagn Res 11(8): BC04

Kunz R (2009) A breakthrough in rice fortification. World Grain 8:58–60

Liu S et al (2011) Extruded moringa leaf–oat flour snacks: physical, nutritional, and sensory properties. Int J Food Prop 14(4):854–869

Ma T, Zuazaga G (1942). Estimation of protein. In: Ranganna, S (ed) Analysis and quality control for fruits and vegetables products. 2nd edn. Tata McGraw Hill Pub. Co., New Delhi, pp 3–10

Mishra A et al (2012) Preparation of rice analogues using extrusion technology. Int J Food Sci Technol 47 (9):1789–1797

Pandiselvam R, Venkatachalam T (2014) Important engineering properties of paddy. J Agric Eng 4:73–83

Sadasivam S, Manickam A (1992) Biochemical methods for agricultural sciences. Wiley eastern limited

Snedecor G, Cochran W (1994) Statistical methods (English edition). Affiliated East-West Press Pvt. Ltd., New Delhi

Tesfaye M et al (2015) Anemia and iron deficiency among school adolescents: burden, severity, and determinant factors in southwest Ethiopia. Adolesc Health Med Ther 6:189

Widara SS, and Budijanto S (2012) Study of rice analogue production from various carbohydrate sources using hot extrusion technology. Institusi Pertanian Bogor

Yogeshwari R, Hemalatha G, Vanniarajan C, Saravanakumar R, Kavithapushpam A (2019) Development of micronutrient fortified extruded rice analogues. Eur J Nutr Food Saf 1–11

Zebeli Q, Aschenbach JR, Tafaj M, Boguhn J, Ametaj BN, Drochner W (2012) Invited review: role of physically effective fiber and estimation of dietary fiber adequacy in high-producing dairy cattle. J Dairy Sci 95(3):1041–1056

Moringa and Its Genome: Future Prospects

15

N. Manikanda Boopathi, M. Raveendran,
and Chittaranjan Kole

Abstract

Moringa oleifera is universally referred as the miracle plant or the tree of life owing to its uses, particularly with respect to medicine and nutrition. Accordingly, there is a huge global market demand for the Moringa products. The use of *M. oleifera* leaf, seed and flower powder in various food supplements has been shown to enrich their nutrient contents. However, many of the studies summarized in this book need further validation to substantiate their findings. The knowledge gap in the Moringa research is provided in this chapter, and possible future studies on Moringa genomics as well as the need for a well-structured and planned experimental designs are also narrated in this chapter.

N. M. Boopathi (✉) · M. Raveendran
Department of Plant Biotechnology, Center for Plant Molecular Biology & Biotechnology, Tamil Nadu Agricultural University, Coimbatore, India
e-mail: nmboopathi@tnau.ac.in

C. Kole
Raja Ramanna Fellow, ICAR-National Institute for Plant Biotechnology, New Delhi, India

15.1 Demand for Moringa Products

The coronavirus catastrophe is having a progressive effect on stimulating demand for immune-boosting supplements and natural products, and such a trend is expected to continue in the future. The main drivers that encourage the trend are increasing life expectancy of consumers, rising disposable incomes and growing consumer awareness. Consequently, health product companies are looking for natural ingredients that give active and functional properties to their products, whilst consumers are looking for natural alternatives to synthetic products.

Dried *M. oleifera* powder is mainly used in nutritional supplements in the European and American health products sector, and it is also recommended for boosting the immune system. *Moringa stenopetala* leaf powder, an African Moringa product, also occupies the European market, but its share remains small.

The European dietary supplements market is projected to reach USD 20.9 billion by 2026. It is also forecasted that the market will grow at a compound annual growth rate of 4.9 percent between 2019 and 2026 (Gonzalez and van der Maden 2015).

Dried Moringa leaf powder has also been shown as source of plant-based protein as it has all essential amino acids. The European plant protein market was worth USD 5.8 billion in

N. M. Boopathi et al. (eds.), *The Moringa Genome*, Compendium of Plant Genomes,
https://doi.org/10.1007/978-3-030-80956-0_15

2018. The market is expected to reach USD 9.5 billion by 2027 (Gonzalez and van der Maden 2015). Owing to the rise in vegetarianism and veganism, European consumers are looking at ways to reduce meat intake and prefer plant-based proteins for their health benefits. The COVID 19 pandemic also has further increased demand for immune-boosting health products, including Moringa value-added products.

Therefore, the highest prospective countries for Moringa products are the Western European countries, including Germany, United Kingdom, France, Netherlands, Italy and Spain, as these countries have the biggest nutritional supplement markets, as well as relatively high awareness on Moringa. Traders and suppliers of Moringa value-added products are also stationed in these countries.

15.2 Production Constraints

India is the primary producer of Moringa in the world, and the major Moringa producing Indian states are Andhra Pradesh, Tamil Nadu and Karnataka. In India, most Moringa is produced on large plantations, and the cultivation practices of Moringa varies between different parts of India. For example, in the Northern and Eastern parts, leaves are harvested in December and January.

Majority of the Southern states of India cultivate Moringa for their tender pods, which impart peculiar aroma to the dishes besides serving as a vegetable. In the Southern parts, the peak season for pod (commonly called as drumsticks) harvesting is between July/September and March/April. However, there is a huge demand for Moringa pods during the lean season (December to February) where there is no fruit set.

Several attempts have been made to increase the off-season production through canopy management (such as pruning) and chemical spray (such as uniconazole @ 50 ppm) to increase the pod yield (Example: Bharathi et al. 2019). However, it has not yet been found widespread among the farmers since most of the farmers are living in resource limited environments.

Moringa is also cultivated in Asia, Africa and Latin America, and in countries such as the Philippines and Bangladesh, Pakistan, South Africa, Ghana, Mozambique, Kenya, Zambia, Pakistan, Brazil, and Peru mainly for leaves. Though leaf production does not require any special attention during the cultivation process, increasing the biomass production to meet the market demand, is the need of the hour.

Furthermore, global Moringa production also experiences the following hurdles, which seriously affects the Moringa leaf or biomass production:

(i) non-availability of suitable cultivars that are adapted to the local environment,

(ii) occasional incidence of pests (for example, budworm (*Noorda Moringae*), leaf caterpillar (*Noorda blitealis*), hairy caterpillar (*Eupterote mollifera*), fruit fly (*Gitonadi stigmata*) and tea mosquito bug (*Helopeltis theivora*) are found to be devastating pests in Southern India), and

(iii) diseases (examples of diseases that occur in Moringa includes brown leaf spot (*Cercospora moringicola*), septoria leaf spot (*Septoria lycopersici* Speg.), alternaria leaf spot (*Alternaria solani* Sorauer), powdery mildew (*Leveillula taurica* Lev. Arn.), root rot (*Diplodia* sp.), fusarium wilt (*Fusarium oxysporium f. sp. Moringae*), fusarium wilt (*Fusarium oxysporium f. sp. Moringae*), fruit rot (*Cochliobolus hawaiiensis* Alcorn), damping off (*Rhizoctonia solani* Kuehn), dieback (*Fusarium semitectum* Berk), anthracnose (*Colletotrichum chlorophyti* Chandra), twig canker (*Fusarium pallidoroseum* Cooke Sacc.) and rust (*Puccinia Moringae* Koorders)).

15.3 Research Gaps that Need to be Filled

This book summarized the topical works in Moringa research in cytology, genetics, breeding, nanotechnology, bioinformatics and health sectors. As indicated in different chapters of this book, we have information on the genetic basis of a range of agronomically and medicinally important traits in Moringa that could provide the basis for genome-scale investigations and the identification of individual genes responsible for the highlighted traits. However, there is a lot more to be done by the Moringa research community. The future research on Moringa may revolve around the following research arena, but not limited to, as there is an increasing demand for this marvellous Moringa products:

1. Though two different draft genome sequence of Moringa were currently released (Tian et al. 2015 and Chang et al. 2019), further improvements of assembly are required in its total length and number of scaffolds to decipher the complete genome sequence. Trends in the long-read sequencing technologies will allow to obtain the whole genome sequence of Moringa at chromosome level. For these studies, long-read (>10 kb) sequencing like PacBio or nanopore sequencing technology is essential (see Chap. 10), which allows to develop large contigs by *de novo* assembling. In addition to improvement of assembly by long sequence reads, several strategies for bridging contigs are also required for finalizing complete genome sequence by developing robust bioinformatics pipelines.
2. In addition to complete genome sequence, development of high-density linkage map will be greatly helpful for finding genes for agronomic traits and that are involved in biosynthesis of medicinal substances. Apart from the above applications, the linkage map based on genetic recombination is also helpful in aligning contigs besides analysing physical structure of the genome. Since it is difficult to identify causal genes for unique characteristics or traits in Moringa from homologs in other plant species, genetic mapping strategy would be greatly beneficial. In point of fact, available genomic and genetic information in Moringa is still insufficient and need to be explored fully.
3. Moreover, upon identification of such candidate genes, their biological functions should be elucidated, for which establishment of more efficient and easier method of Moringa genetic transformation is required, which is not yet attempted so far.
4. An advanced understanding of all the developmental and biochemical processes involved in the expression of key Moringa phenotypes (such as increased leaf biomass, reduced flower drops and enhanced fruit set) is warranted. On the other hand, such progress in both breeding and physiology areas relies strongly on detailed examinations of the genes and resolution of their molecular and metabolic regulatory circuits. Although the molecular tools are available and the key concepts are established in the model as well as field crops, the key issue here is the availability of sufficient funds and research power to confront these challenges in Moringa.
5. Though there are considerable amount of scientific information on Moringa production technologies and value addition to the Moringa products (Chap. 14), the genetic improvement of Moringa for pest and diseases are still in its infant stage. The regular breeding strategies such as hybridization and backcross strategies are found to be a herculean task in Moringa since it is a tree crop and having fragile stem. As an alternative strategy, the recent developments in biotechnology arena, such as use of Moringa genomic information in mutagenesis and targeted induced local lesions in genome (TILLING; Gao et al. 2020) and genome editing (including clusters of regularly interspaced short palindromic repeats (CRISPR), CRISPR-associated 9 (Cas9) proteins; Osakabe et al. 2020), can be employed since there is no such

effort so far reported in Moringa. Other areas that have potential to offer rich genetic information but not yet to be practised in Moringa are Quantitative Trait Loci (QTL) mapping and Marker-Assisted Selection (MAS) and Genome-wide association studies (GWAS). To introgress a trait from the donors, the only progeny retained are those containing markers flanking the locus of interest from the donor parent. This speeds up and reduces the expense of growing and phenotyping large numbers of progenies which will greatly enhance the speed of Moringa genetic improvement programme. In addition, if the causative genetic changes can be identified, then it is possible to transfer such information across plant species. It is sincerely hoped that as the Moringa genome sequence is available, we may see a burst of research papers identifying causal genes for a range of phenotypes in Moringa in near future.

6. Though there are 12 related Moringa species with desirable traits (such as adaptive alleles that confer drought stress, crop protection from pest and diseases and unique medicinal properties from Moringa wild relatives), almost no attempt has been undertaken to introgress such traits from wild relatives to *M. oleifera*. It is strongly believed that the recent developments in high-throughput molecular marker technologies and improved marker-assisted selection strategies would increase the efficiency of introgression of the desirable alleles from the locally adapted ecotypes and wild relatives to the elite Moringa cultivars. Such research should also focus on understanding the genetic relationships among the species since all wild species are not cross-compatible with *M. oleifera*. This also warrants extensive Moringa germplasm collection, conservation and characterization at all possible ways (*i.e.* at morphological, biochemical, molecular and pharmacological level), and make the information readily available to the research folks through community approaches. Examinations of wild relatives will also provide data that help to understand ecological and evolutionary phenomena and the contribution of phenotypic plasticity of ecotypes of Moringa to evolution.
7. There is also research gap in Moringa-based production and processing technologies as furnished below:
   a. Development of affordable and efficient strategy that favours off-season Moringa pod production as it always has huge market potential for pods during lean season (December–February in India).
   b. Increasing the shelf-life of Moringa fresh leaves and pods as it withers gradually after harvest and reduce the market price.
   c. Evolving efficient strategies (such as precision farming, protected cultivation and hydroponics) to cultivate Moringa in resource unlimited urban areas.
8. Production of unique medicinally imperative metabolites or functional ingredients characterizes Moringa as an exceptional medicinal plant. Though biochemical studies were employed for elucidating its physiological functions, employing genetic or molecular genetic studies are limited. Therefore, exploring such approaches in analysing the synthesis of functional metabolites or proteins will certainly contribute to dissect their biosynthesis pathways or to find their regulatory genes.
9. Mehwish et al. (2020) reported that *M. oleifera* can used as a potent immuno-modulator and when ingested could enhance the both cell-mediated and humoral immunity of host. However, further collaborations between chemists, biologists, physicists, industrialists, clinicians and epidemiologists will be needed to understand the influence of *M. oleifera* on the host immune system.
10. It is also described that *M. oleifera* contributes towards the host health through restructuring the gut microbiota, which in turn helps to control various diseases such as inflammation and to enhance host immunity (Jaja-Chimedza et al. 2018). However, a huge gap in knowledge is clearly noticeable, which

suggests that further in vitro and in vivo investigations are needed to document the immunomodulatory potential of *M. oleifera* supplemented functional foods.

11. Development of Moringa raw products into higher-value products appear to be feasible. However, discovering the market opportunities and meeting the demands of end users seem to be a recurring problem when setting up a Moringa business/project. This gap may be addressed with suitable systematic strategies.
12. Social scientists have a big role to play in Moringa cultivation and consumption and some of them are highlighted hereunder:
    a. Novel strategies, tools and gadgets have to be developed to promote awareness around the nutritional value and market potential of Moringa products in order to maximize the returns of Moringa trees.
    b. A detailed cost and benefit analysis of Moringa production should be conducted.
    c. Moringa suppliers or company stakeholders need to get acquainted with the regulations and standards required when targeting the export market. This also means that suppliers should establish, and nurture, trading relationships with importers or even intermediaries since the volume supplied is likely to be limited according to each country's terms.
    d. Studies should also be initiated to provide an account of the legislative and non-legislative requirements to comply with and/or needs to be followed by Moringa producers in developing countries.

Nevertheless, research so far have demonstrated that Moringa is not only worth to study as it is a nutritionally dense, medicinally and economically imperative crop but it is also an excellent model to study ecological and evolutionary processes as it is adapted to varying climatic conditions starting from the hilly slopes to the dry sandy regions. Though undertaking aforementioned methods and strategies, to fill the research gap, needs a lot of effort, resources and time for growing and screening a large number of breeding materials. Definitely such efforts would be beneficial to evolve novel Moringa cultivars with improved health benefits besides to ensure food, nutritional and fiscal security since they promote both basic and applied scientific studies in Moringa.

References

Bharathi CS, Pugalendhi L, Mohan B (2019) Effect of canopy management and chemical manipulation on morphological attributes of off-season production in annual Moringa cv. PKM1. Int J Curr Microbiol Appl Sci 8(2):2642–2648

Chang Y, Liu H, Liu M, Liao X, Sahu SK, Fu Y, Hendre PS (2019) The draft genomes of five agriculturally important African orphan crops. GigaScience 8 (3):giy152

Gao YL, Yao XF, Li WZ, Song ZB, Wang BW, Wu YP, Liu CM (2020) An efficient TILLING platform for cultivated tobacco. J Integr Plant Biol 62(2):165–180

Gonzalez YS, van der Maden ECLJ (2015) Opportunities for development of the Moringa sector in Bangladesh: desk-based review of the Moringa value chains in developing countries and end-markets in Europe (No. 15-102). Centre for Development Innovation, Wageningen, UR

Jaja-Chimedza A, Zhang L, Wolff K, Graf BL, Kuhn P, Moskal K, Raskin I (2018) A dietary isothiocyanate-enriched Moringa (*Moringa oleifera*) seed extract improves glucose tolerance in a high-fat-diet mouse model and modulates the gut microbiome. J Funct Foods 47:376–385

Mehwish HM, Riaz Rajoka MS, Xiong Y, Zheng K, Xiao H, Anjin T, He Z (2020) *Moringa oleifera*–a functional food and its potential immunomodulatory effects. Food Rev Int 1–20

Osakabe K, Wada N, Miyaji T, Murakami E, Marui K, Ueta R, Osakabe Y (2020) Genome editing in plants using CRISPR type ID nuclease. Commun Biol 3 (1):1–10

Tian Y, Zeng Y, Zhang J, Yang C, Yan L, Wang X, Huan YZ (2015) High quality reference genome of drumstick tree (*Moringa oleifera* Lam.), a potential perennial crop. Sci China Life Sci 58(7):627–638

The manufacturer's authorised representative in the EU is Springer
Nature Customer Service Centre GmbH, Europaplatz 3, 69115 Heidelberg,
Germany. If you have any concerns regarding our products, please
contact ProductSafety@springernature.com

Printed and bound by CPI Group (UK) Ltd, Croydon, CR0 4YY
30/09/2026
02225687-0016